MEDIUM ÆVUM MONOGRAPHS

EDITORIAL COMMITTEE

K. P. CLARKE, A. J. LAPPIN,
N. F. PALMER, P. RUSSELL, C. SAUNDERS

MEDIUM ÆVUM MONOGRAPHS

ON LIGHT

EDITED BY

K. P. CLARKE & SARAH BACCIANTI

The Society for the Study of Medieval Languages and Literature
OXFORD · MMXIV

THE SOCIETY FOR THE STUDY OF
MEDIEVAL LANGUAGES AND LITERATURE

OXFORD, 2014

http://mediumaevum.modhist.ox.ac.uk

© The Authors, 2014

British Library Cataloguing in Publication Data
A catalogue record for this book is available from
the British Library

ISBN-13: 978-0-907570-29-5 (pb)

CONTENTS

List of Illustrations   vii

K. P. Clarke and Sarah Baccianti   1
    Introducing Light

Eric Gerald Stanley   5
    Light for Oxford

Michael J. Huxtable   25
    The Relationship of Light and Colour in Medieval
    Thought and Imagination

David M. Barbee   45
    The Utility of the *Lux-Lumen* Distinction in Roger
    Bacon's Thought

William T. Rossiter   63
    'The light so in my face | Bigan to smyte':
    Illuminating Lydgate's *Temple of Glas*

Hannah Hunt   87
    Divine Light and Spiritual Intoxication: Symeon the
    New Theologian's Image of Penitence as a Mystical
    Winepress

Cecilia A. Hatt   107
    Lux mediatrix: 'All that is made manifest is light'
    (Ephesians 5. 13): The Influence of Robert
    Grosseteste's Theory of Light on Bishop John Fisher

Virginia Langum   121
    Seeing in Sermons: Word, Light and Aesthetic Experience

Joy Hawkins   137
    Sights for Sore Eyes: Vision and Health in Medieval England

Sabina Zonno   157
    Illuminated Darkness: The Image of the Blind or Blindfold Man in Some Thirteenth- and Fourteenth-Century European Manuscripts

Stephanie Seavers and Catia Viegas Wesolowska   177
    Light and Virtue: The Gloucester Candlestick

Abbreviations   191

Bibliography   192

LIST OF ILLUSTRATIONS

following p. 176

Zonno, Fig. 1. Brescia, Biblioteca Queriniana, MS A. V. 17, fol. 7^r.
Zonno, Fig. 2. London, British Library, MS Yates Thompson 13, fol. 103^r.
Zonno, Fig. 3. Padova, Biblioteca Capitolare, MS B. 14, fol. 177^r.
Zonno, Fig. 4. Oxford, Bodleian Library, MS Bodley 264, fol. 74^v.

following p. 190

Seavers and Viegas, Fig. 1. The Gloucester Candlestick.
Seavers and Viegas, Fig. 2. Finished sample casting using silver copper alloy.
Seavers and Viegas, Fig. 3. Detail of the Knop.
Seavers and Viegas, Fig. 4. Detail of the base.
Seavers and Viegas, Fig. 5. Detail of beneath the drip pan.

Introducing Light

K. P. Clarke and Sarah Baccianti

It is impossible to do without light. This much is clear to anyone engaging with the intellectual breadth and creative scope of the Middle Ages and its literary, theological, scientific and material culture. The essays offered here have not been gathered with the aim of exhaustively representing this breadth and scope but they do give a good sense of how interdependent these fields are, how subject to an exciting cross-fertilization. *On Light* opens with Eric Stanley's 'Light for Oxford', a wide survey of literary, biblical uses of the figure of light, beginning with what he wryly dubs the 'Oxford Psalm', *Dominus illuminatio mea* (Psalm 27. 1). With minute philological attention to the words for light, especially in Latin and Old English, Stanley traces their varied use (often alongside darkness), ending with the comic interdependence of light and dark in Chaucer's *Reeve's Tale*. 'If only', he says, 'on the first day of Creation, God the All-knowing had foreseen the fracas at the mill in Trumpington, he might have divided day from night, the light from the darkness, less severely'.

Michael Huxtable's essay, 'The Relationship of Light and Colour in Medieval Thought and Imagination' highlights how crucial colour is in medieval theoretical approaches to light and how there are, consequently, important epistemological dimensions to be considered. 'Colour is', he says, 'the limit of the translucent power in a determinately bounded body', 'the external edge of the material nature of a physical body'. To *know* colour, as well as its relationship to light, is to have a heightened, deeper understanding of the physical world around us. The use to which colour was put, specifically in the fourteenth century, is explored further by Huxtable with tantalising further work promised on how these inflect the *green*ness of the Green Knight encountered by Sir Gawain.

David Barbee takes one of the central distinctions in medieval theories of light, inherited Greek natural philosophy, between *lux* and *lumen* and shows how vital the distinction was as an organizing principle in the work of Roger Bacon. The metaphor of light was used to explain his illumination theory of human knowledge; God's creation of the universe was similar to the way light extends from light; and light as a model for motion allowed him to prove the existence of God as first cause. As Barbee says, 'The *lux-lumen* distinction, then, would seem to cut across disciplinary boundaries as a valid explanatory model based on the simple fact that the categories spiritual and corporeal are universally inclusive.'

William T. Rossiter looks at the use of light in Lydgate's *Temple of Glas* focussing in particular on how the poem relates to various contemporary discourses on light, vision and reflection. Arguing for Lydgate's sensitivity to scientific theories of light, Rossiter suggests that poetry and science are in dialogue in the poem and that Lydgate is deploying multiple modes and registers in the *Temple*.

The moral uses to which light might be put are explored in three essays by Hannah Hunt, 'Divine Light and Spiritual Intoxication: Symeon the New Theologian's Image of Penitence as a Mystical Winepress'; Cecilia Hatt, '*Lux mediatrix*: 'all that is made manifest in light' (Ephesians 5. 13): The Influence of Robert Grosseteste's Theory of Light on Bishop John Fisher'; and Virginia Langum, 'Seeing in Sermons: Word, Light and Aesthetic Experience'. Hunt shows how pervasive light imagery is in the sermons of St Symeon and concentrates in particular on *Catechesis* 23, a striking example of Symeon's skill at fusing theological complexity and literary eloquence. Hatt examines the way that Fisher, in a sermon on light, establishes the Virgin Mary as implicated in a daily, wonderful, but entirely natural phenomenon. 'The light', she asserts, 'of which [the Virgin Mary] is privileged to be a vehicle is an epistemic agent, both object and means of knowledge, and is not really metaphorical at all.' Langum's essay, too, shares this sense of metaphor being insufficient in accounting for the uses of light imagery in medieval sermons. Analogous to sight having both an extramissive (active) and intromissive (passive) aspect, Langum borrows the term 'active discovery' from the aesthetic theory of Monroe Beardsley in order to think about the active involvement and 'inward seeing' of the listener to a medieval sermon. The spiritual and physical causes of blindness stand in parallel: for example, pride to smoke; avarice to shiny metal; sloth to smoky vapours. The spiritual is correlative to everyday, familiar, phyiscal phenomena. It is real.

The physical experience of light is inextricably bound up with the sense of sight, pre-eminent among the senses. The relationship between seer and seen was effectively one of touch, since sight was a complex interaction of intromission and extramission. Two essays directly engage with sight: Joy Hawkins, 'Sights for Sore Eyes: Vision and Health in

Medieval England' and Sabina Zonno, 'Blind or Blindfold: The Image of the Blind Man in Some Fourteenth-Century European Illuminations'. Hawkins looks at the preponderance of cures of eye complaints in medieval herbals, emphasising how sight had a vital effect on the whole physiological system. The deterioration of sight had serious consequences: one was no longer able to participate fully in the Mass, for example, by witnessing the raising of the Host. Sabina Zonno explores a variety of representations of blindness in medieval psalters, books of hours, bibles, as well as illustrations of games involving the blind or blindfolded. Many of these representations are nourished by the numerous biblical connections between sin and blindness, seeing lack of sight as the physical manifestation of a moral state.

The reality of light, its physical materiality drives the final essay, by Stephanie Seavers and Catia Viegas, 'Light and Virtue: The Gloucester Candlestick'. Making sense of this fascinating, beautiful object requires an attention to many of the fields explored in the essays above, science, theology, literature to supplement and inflect the detailed technical examination carried out on it. They discuss how the candlestick was made, as well as how to interpret its inscriptions, arguing that the rich symbolism around light in the middle ages is fully reflected in the candlestick's construction. Discussing the way the candlestick was cast, using the 'lost wax' method, where a piece is constructed with gaps or holes, they suggestively remark that in many ways the candlestick was not just made to give light, but was made *with* light.

Like the image on the cover of this volume, the roles of light and darkness cover both art and literature, from candlesticks to miniature, from architecture to religion, from science to poetry. As Joan Tasker Grimber writes, the contrasts that are present in medieval literature, especially in poetic works, are underlain by the *clair-obscur*, 'by the dialectic of light and darkness, of day and night, [... as] love of light and fear of darkness is so universal that it surely must be considered as one of the archetypes of human thought'.[1] The range of articles in this collection offers an overview of this eclectic use of *clair-obscur*, providing different points of discussion on the notion of physical and mystical light, on blindness and art and on morality. Light is, in the words of Attilio Mellone (here with specific reference to Dante) 'una specie d'idea

---

[1] Joan Tasker Grimbert, 'Effects of *Clair-Obscur* in *Le Bel Inconnu*', in *Courtly Literature: Culture and Context*, ed Keith Busby and Erik Kooper (Amsterdam: J. Benjamins 1990), pp. 249–60 (p. 250).

matrice, di filo conduttore' ('a kind of constant idea in the background, a central theme').[2]

The essays gathered in this volume comprise the selected papers delivered at the fourth Oxford Graduate Medieval Conference at Lincoln College, in April 2008. The editors gratefully acknowledge the financial assistance of the Society for the Study of Medieval Languages and Literature, the Faculty of English Language and Literature, Oxford, and the Rector of Lincoln College. Their commitment to medieval studies is matched only by the commitment and enthusiasm of the medieval graduate community at Oxford and the conference's delegates and participants.

---

[2] Mellone is here discussing the use of light in the work of Dante, see *Enciclopedia dantesca*, gen. ed. Umberto Bosco, 6 vols (Rome: Istituto delle Enciclopedia italiana, fondata da Giovanni Treccani, 1970–1976; 2nd edn 1984), III, p. 708.

LIGHT FOR OXFORD

Eric Gerald Stanley

The Oxford Psalm

It may seem supererogatory to have a conference on LIGHT in Oxford. Like carrying coals to Newcastle, or better, one might echo the question, perhaps already proverbial, asked in *The Birds* of Aristophanes, 'Who has brought this owl to Athens?', the city whose coins were adorned with the image of an owl.[3] In Oxford light is emblazoned in the crest of the University: *Dominus illuminatio mea*. That is the opening of Psalm XXVI in the numbering of the Vulgate, rendered in the Doway Version, in which, when possible, Latinisms corresponding to the Vulgate wording are preferred to kersey English, 'OVR Lord is my illumination.'[4] A year later the Authorized Version was published, and in it the Hebrew of Psalm XXVII verse 1, is translated, 'THe LORD *is* my light, and my saluation, whome shal I feare? the LORD *is* the strength of my life, of who*m* shall I be afraid?'[5] A learned, but now somewhat antiquated nineteenth-century commentary tells us that in sound the Hebrew for 'my light' is close to the sound of the Hebrew for 'shall I be afraid?'[6] There is paronomasia at the centre of the verse. By wordplay the Psalmist has brought light and fear together. In view of what follows on Holy Saturday in this paper, it deserves to be remembered that this psalm is said as the priests walk in the Easter Vigil procession with unlighted

---

[3] Jeffrey Henderson, ed. and trans., *Aristophanes*, III, Loeb Classical Library (Cambridge, Massachusetts: Harvard University Press, 2000), pp. 58, 301.

[4] *The Holie Bible Faithfully Translated into English ... By the English College of Doway*, 2 vols (Doway: by Laurence Kellam, 1609, 1610), vol. II, p. 56.

[5] *The Holy Bible, An Exact Reprint in Roman Type ... of the Authorized Version Published in the Year 1611*, intro. Alfred W. Pollard (London: Henry Frowde, Oxford University Press, 1911), unpaginated.

[6] J. M. Neale, and R. F. Littledale, *A Commentary on the Psalms from Primitive and Mediæval Writers*, 4 vols (4th edn; London: Joseph Masters & Co., 1884), vol. I, p. 373. None of the more recent commentaries I have consulted refers to this wordplay.

candles (according to the *Sarum Missal*).⁷ The lighting of the Paschal candle comes later in the rites of that festival. To avoid specifying the number of the psalm, whether Vulgate or Authorized Version, I shall call it 'the Oxford Psalm', not without some institutional self-aggrandizement.

'Good night!' in several languages

Modern lighting, first by gas, then by electricity, has dispelled darkness, and much of the fear that went with it. Security and hope come with light, and some sense of joyousness. It did so in the eighteenth century, when less luminous artificial light was carried in by servants at court and in the palaces and houses of the great and prosperous. Goethe travelled in Italy, and he records how in Venice, on 5 October 1786, he went to the theatre. On his way back to where he was staying, he had still ringing in his ears the cries of the audience, *bravo, bravi*, and thought of the difference between what he had just heard in Venice and what he would have heard in Germany when people say goodbye to each other after a visit to the theatre:⁸

> Gute Nacht! so können wir Nordländer zu jeder Stunde sagen, wenn wir im Finstern scheiden, der Italiäner sagt: *Felicissima notte!* nur einmal, und zwar wenn das Licht in das Zimmer gebracht wird, in dem Tag und Nacht sich scheiden, und da heißt es denn etwas ganz anderes. So unübersetzlich sind die Eigenheiten jeder Sprache: denn vom höchsten bis zum tiefsten Wort bezieht sich alles auf Eigenthümlichkeiten der Nation, es sey nun in Charakter, Gesinnungen oder Zuständen.

> ['Gute Nacht!' So we Northerners can say at any hour whenever we part in the dark; the Italians say *Felicissima notte!* once only, namely when lights are brought into the room as day parts from night, and so it means something

---

⁷ J. Wickham Legg, ed., *The Sarum Missal* (Oxford: Clarendon Press, 1916, repr. 1969), p. 115.

⁸ Johann Wolfgang von Goethe, *Goethe's Werke: Vollständige Ausgabe letzter Hand*, 60 vols (Stuttgart & Tübingen: Cotta, 1828–1842), *Italiänische Reise*, I, vol. XXVII [1829], p. 125. I have used this quotation in an earlier article to illustrate the indeterminacy of translation: see E. G. Stanley, 'Polysemy and Synonymy and How these Concepts were Understood from the Eighteenth Century Onwards in Treatises, and Applied in Dictionaries of English', in Julie Coleman and Anne McDermott, eds, *Historical Dictionaries and Historical Dictionary Research*, Lexicographica Series Maior, 123 (Tübingen: Max Niemeyer, 2004), pp. 157–83 (pp. 159–60). Throughout this paper I depart from editorial details in quoting texts. All translations, unless otherwise stated, are mine.

quite different. Thus the idiotisms of every language are untranslatable; for from the highest to the lowest word everything relates to the peculiarities of the country, whether in its character, or in its sentiments, or in its circumstances.]

Goethe was expecting to hear from the Venetian theatre-goers 'Good night' as they walked home from the theatre, through dark and dangerous streets, or to their gondola to carry them to safety. The greetings, superficially so near in sense, were not alike in occasion in the two languages.

The Paschal

The darkness of a sad world is illumined by joyous anticipation of the Day of Resurrection when on Holy Saturday, Easter Day almost, the Paschal candle is lighted, accompanied by the singing of the *Exultet*. The great South Italian churches of the early Middle Ages have their ornate Paschal candle-stick, the baldachin over the altar at the centre of the chancel, and the bishop's throne at the East end. Two ornate marble ambones, rectangular pulpits of splendour, begin the nave, and from one of them was hung an *Exultet* roll, gloriously ornamental and festive, showing upside down some of the words of that prayer and the neums that are part of it, and the illuminations of the prayer the right way up for the congregation to see. The Sarum Missal has the Latin words as used in England:[9]

> Exultet iam angelica turba celorum exultent diuina misteria et pro tanti regis uictoria tuba intonet salutaris. Gaudeat se tellus tantis irradiata fulgoribus et eterni regis splendore lustrata tocius orbis se senciat amisisse caliginem. Letetur et mater ecclesia tanti luminis adornata fulgoribus et magnis populorum uocibus hec aula resultet. Quapropter astantibus uobis fratres karissimi ad tam miram sancti huius luminis claritatem una mecum queso dei omnipotentis misericordiam inuocate. Ut qui me non meis meritis intra leuitarum numerum aggregare dignatus est luminis sui gracia infundente cerei huius laudem implere perficiat.

In F. E. Warren's translation:[10]

---

[9] The Sarum wording is at Legg, ed., *The Sarum Missal*, p. 118.
[10] Frederick E. Warren, trans., *The Sarum Missal in English*, 2 vols, The Library of Liturgiology & Ecclesiology for English Readers, ed. Vernon Staley, VIII, IX (London: The De La More Press, 1911), vol. I, pp. 270–71. What I say about these rolls is entirely based on the plates in Myrtilla Avery, ed., *The Exultet Rolls of South*

> Now let the angelic host of heaven exult, let the divine mysteries be celebrated with exultation, and let the trumpet of salvation sound for the victory of so great a King.
>
> Let the earth rejoice, irradiated with so great brilliancy, and illuminated by the splendour of the eternal King, let it perceive the darkness of the universe to have been done away.
>
> Let mother church also be joyful, adorned with the brilliancy of so great a light, and let this court[11] resound with the mighty voices of peoples.
>
> Wherefore, most dearly beloved brethren, as ye stand before the so wonderful brilliance of this holy light, I beseech you invoke along with me the tender mercy of almighty God;
>
> That he who hath deigned to enrol me, not for mine own merits, within the number of the Levites, pouring forth upon me the grace of his light, may cause me to declare fully the praise of this taper.

Warren translates the details of the ceremonies, how the unlighted candles are brought in, how the Paschal candle is lighted, and how it shall burn continuously until Ascension Day, forty days after Easter; and how towards the end of the prayers for Holy Saturday candles throughout the church are to be lighted, dispelling the darkness.[12] He gives details of the *Vetus Itala* version in which the bees that produced the wax for the Paschal candle are honoured at greater length, as splendidly shown in the illuminations of some of the *Exultet* rolls.

Anglo-Saxon England and southern Italy

It might seem that all this about churches in southern Italy and illuminations on *Exultet* rolls has nothing much to do with England; after all, it is more than 2000 km from somewhere in northern England to Subiaco, St Benedict's first monastery, and Montecassino, his second monastery from where the Benedictines radiated throughout learned Europe including far off England. However, it took the Anglo-Saxons no

---

Italy, II (all published) Plates (Princeton: University Press, 1936), 55 + CCVI plates, and it is regrettable that vol. I was never published, since only selections of the text are reproduced. The wording of the Vetus Itala, as used in the earliest rolls, differs from the Sarum text in some details.

[11] Sarum *aula*; in Medieval Latin *aula* often means 'church, nave'.
[12] Warren, trans., *The Sarum Missal in English*, pp. 270–73. Legg, ed., *The Sarum Missal*, pp. 117–18, quotes much of these ceremonial details from Paris, Bibliothèque de l'Arsenal, MS 135. The early manuscripts show the bees in several of Avery's sets of plates.

longer to reach Rome from England, it took the Benedictines no longer to reach Monkwearmouth and Jarrow from Montecassino, than it would have taken anyone till the middle of the nineteenth century when the railways opened Europe to people carrying Baedekers instead of Bibles, and they began to echo the Italians and to say that Christ stopped at Eboli, south of Naples. Perhaps he did, but perhaps Christ came from the South.

What all this leads up to is the fact that the great pages in the Lindisfarne Gospels showing the Evangelists have affinities with manuscripts thought to emanate from the home of Cassiodorus and the monastery he founded in the middle of the sixth century on his estate at Vivarium (Squillace, a few kilometres south of Catanzaro in southern Calabria), almost as far south as you can go on the mainland of Italy. For the tracing of such affinities one is dependent on art historians, learned, but, when dealing with an age of which so much has been lost to posterity, inevitably speculative when drawing conclusions from the evidence we and they now have. The art of the great books of Northumbria, the Codex Amiatinus with its portrait of Ezra, presented like Cassiodorus, and the Lindisfarne Gospels with their portraits of the Evangelists, is thought to go back ultimately to the Vivarium, not far short of 3000 km south of Monkwearmouth and Jarrow.[13] The art of Northumbria around the time of Bede consists not only in beautiful writing, but also in textual accuracy, for the wording of the Northumbrian Codex Amiatinus, now in Florence, underlies the version of the Vulgate authorized by Rome. The art lies most strikingly in illumination, the very word used by the Doway translators of the Oxford Psalm. Some important parts of that illumination draw on the art practised in farthest Calabria.

Mirk's Festial on the Paschal and on Ascension Day

Much later in the Middle Ages John Mirk's *Festial*, written in the 1380s, can be relied upon to give us the commonplaces for the Christian year.

---

[13] See Rupert Bruce-Mitford, in T. D. Kendrick, T. J. Brown, R. L. S. Bruce-Mitford et al., eds, *Evangeliorum Quattuor Codex Lindisfarnensis*, 2 vols (Olten and Lausanne: Urs Graf, 1956, 1960), vol. II, pp. 142–57, chapter 3, 'The Sources of the Evangelists' Portraits'.

What he says of the Paschal, that is, of Holy Saturday, relates the Paschal, that is, the Paschal candle, to the Pillar of Fire that guided the Children of Israel to the Promised Land, and, echoing the Oxford Psalm and, as we shall see, echoing the exegetes on words in the Sermon on the Mount, he conveys the spiritual meaning of the Paschal candle: its light signifies Christ:

> On Astyr Euen þe Paschall is makyþe, þat bytokenyþe Crist: for as þe Paschall is þe chef tapor þat is in þe chyrch, so is Crist þe chef saynt þat ys in þe chyrch. Also þys Paschall bytokeneþe þe pyler of fure þat ʒode befor þe Chyldyr of Israell, when þay went out of Egypte into þe Lond of Behest, þat is now Ierusalem. [...] Thus is þe Paschall halowed and lyʒt with new fure, and of hyt all oþyr tapyrs byn lyʒt; for all lyʒt and holynes of good worchyng comyþe of Cristis lore, and Holy Chyrch ys liʒt wyth brennyng charyte of his behestys.[14]

> [On the Eve of Easter the Paschal is made, which signifies Christ: for as the Paschal is the chief candle that is in the church, so is Christ the chief saint that is in the Church. Likewise this Paschal signifies the Pillar of Fire that went before the Children of Israel when they went out of Egypt into the Promised Land, that is now Jerusalem. [...] Thus is the Paschal sanctified and lighted with new fire, and from it all other candles are lighted; for all light and sanctity of good deeds come from Christ's teaching, and Holy Church is alight with the burning Love of his and its commands.]

Mirk takes up the symbolism of the Paschal candle at the beginning of his brief sermon on Ascension Day:

> Good men and woymen, þys day ys an heʒ day, and an hegh fest in all holy chyrche; for þis day, as þe fayþe of Holy Chyrche beleueth and precheþ, Crist, God of Heuen, veray God and man, stegh vp ynto Heuen, and syttyþ þer on hys Fadyr ryght hond in þe blis þat euer schall last. Wherfor yn tokenyng of þys þyng, þat ys þe schef lyght þat ys yn Holy Chyrche, þat haþe stonden from Astyr hedyrto oponly yn þe quere, þys day is remuet away in schewyng þat Crist, þe whech ys þe chef lyʒt yn Holy Chyrch and haþe þes fourty dayes oponly apperyd to hys Dyscyplys by mony wayes, and taʒt hom þe fayþe, thys day he steʒ vp ynto Heuen, and þer schall abyde tyll þe Day of Dome.

---

[14] Thomas Erbe, ed., *Mirk's Festial*, EETS, e.s. 96 (London, 1905), pp. 127–8, 151–52. My paper was last modified in November 2008, using Erbe's text. A year later the now standard edition appeared: Susan Powell, ed., *John Mirk's 'Festial' edited from British Library MS Cotton Claudius A. II*, vol. I, EETS, o.s. 334 (2009); in this edition the corresponding passages are at pp. 112-13, 140.

[Good men and women, today is a high day, and a high festival in all Holy Church; for today, as the Faith of Holy Church confides and preaches, Christ, God of Heaven, very God and man, ascended into Heaven, and sits there on his Father's right hand in the bliss that shall last forever. For which reason, in signification of this thing, which is the chief light that is in Holy Church, that has stood in the choir openly from Easter until now, (this thing) is today taken away, signifying that Christ, who is the chief light in Holy Church and has openly appeared to his Disciples these forty days by many means, and has taught them the Faith, today he ascended into Heaven, and there he shall dwell till the Day of Judgement.]

The semantics of Old English leoht, and of Modern English light

The distance is great from the symbolism of the Paschal candle to the semantics of the Old English noun *leoht*, the etymon of Modern English *light*. The meaning of that word in the vernacular poetry of the Anglo-Saxons is my next subject. As usual, Christian W. M. Grein's dictionary, publication of which was complete by 1864, remains a valuable guide.[15]

Old English poetry is rich in the semantics of the word *leoht*. How manifold the senses are in later English is sufficiently adumbrated by a summary of many of the senses of the noun *light* listed in *The New English Dictionary*, this part, *Lief* to *Lock*, published in 1903 under the editorship of Henry Bradley:[16]

> **1.** That natural agent or influence which (emanating from the sun, bodies intensely heated or burning, and various other sources) evokes the functional activity of the organ of sight. **a.** Viewed as the medium of visual perception generally. Also, the condition of space in which light is present, and in which therefore vision is possible. Opposed to *darkness*. **b.** Viewed as being itself an object of perception, cognized by means of the specific visual sensation

[15] C. W. M. Grein, ed., *Sprachschatz der angelsächsischen Dichter*, 2 vols, Bibliothek der angelsächsischen Poesie, III, 1 and 2 (Cassel and Göttingen: Georg H. Wigand, 1861, 1864), vol. II, pp. 177–78. It was slightly improved in the revised edition by J. J. Köhler and F. Holthausen (Heidelberg: Carl Winter: 1912–1914), p. 417, who supplied some words and senses that had been omitted, and corrected some errors, the result of misreadings of texts for which Grein was hardly responsible, for he himself never saw a manuscript and for all transcriptions had to rely on the work of others.

[16] For details of the publication of *The New English Dictionary* (now *The Oxford English Dictionary*) see Darrell R. Raymond, ed., *Dispatches from the Front: The Prefaces to the Oxford English Dictionary* (Waterloo, Ontario: Centre for the New Oxford English Dictionary, University of Waterloo, 1987). Such derivates as *lightness, lightsomeness*, are ignored in my summary, and I confine myself to senses involving 'luminosity' and some metaphorical extensions of that sense.

indicated by the use of words like 'bright', 'shining', etc. Also in particular sense, an individual shining or appearance of light. **c.** Viewed as residing in or emanating from a luminary. **d.** In scientific use. **e.** The portion or quantity of light which comes through a window, or which is otherwise regulated so as to illuminate a given space. **f.** *In light*: exposed to rays of light, lighted up. **g.** *One's light*: the ordinary measure of light which a person enjoys, or expects to enjoy, for seeing around him. **h.** A gleam or sparkle in the eye, expressive of animated feeling or the like. **i.** *To put out* or *quench* (one's *light*: to extinguish his 'vital spark'. **j.** *pl.* [after L. *lumina*.] Graces of style. **k.** fig. *light of one's eye(s*: applied to a loved object. **l.** *The light of God's countenance*: in Ps. iv. 6, etc. = Divine favour. In allusion to this, *the light of (a person's) countenance* is often sarcastically used for: (his) sanction, approving presence. **2.** *spec.* The illumination which proceeds from the sun in day-time; daylight. Also the time of daylight: day-time, day-break. **b.** In the asseverative phrase *by this (good) light*. Also *by God's light*. **c.** *To see the light*, to come into the world; to be brought forth or published. **3.** The state of being visible or exposed to view, *To come to light*: to be revealed, disclosed, made visible or made known. *To bring to light* (cf. F. *mettre en lumière*): to reveal, make known, publish. **4.** Power of vision, eyesight (now *poet.* or *rhet.*). Also *pl.* = the eyes (now only *slang*). **5.** A body which emits illuminating rays. a. The sun or other heavenly body (after Gen. i. 16). **b.** An ignited candle, lamp, gas-jet, or the like. Hence *wax lights* = wax candles for lighting (now *rare* in this use). **c.** *collect.* The candles or other illuminants used to light a particular place; lights collectively. † Also material to be burnt for lighting. **d.** A signal-fire or beacon-lamp, esp. on a ship or in a lighthouse; often with prefixed qualification as *fixed, flashing, intermittent, revolving light*. Hence, used for the lighthouse itself. **e.** A linkman. *Obs.* **6.** Used *fig.* With reference to mental illumination or elucidation. **a.** In phrases, as *to give light*. Also *to get or receive light*. Now usually *to throw (cast, shed) light upon*. **b.** Illumination or enlightenment, as a possession of the mind, or as derivable from some particular source. *Light of nature*, the capacity given to man of discerning certain divine truths without the help of revelation. **c.** *pl.* (*a*) Pieces of information or instruction; facts, discoveries, or suggestions which explain a subject. (*b*) The opinions, information, and capacities, natural or acquired, of an individual intellect. **d.** *New light(s*: novel doctrines (esp. theological and ecclesiastical) the partisans of which lay claim to superior enlightenment; hence by antithesis *Old light(s*, the traditional doctrines to which the 'new lights' are opposed. **e.** A suggestion or help to the solution of a problem or enigma. **7.** Often with spiritual reference (said of the brightness of Heaven, the illumination of the soul by divine truth or love, etc.). **b.** *spec.* Among Quakers, the inward revelation of Christ in the soul. **c.** Applied to God as the source of divine light, and to men who manifest it. **8.** In figurative uses of sense 5: **a.** One who is eminent or conspicuous for virtue, intellect, or other excellence; a luminary. **b.** A bright example. **9.** In figurative uses of sense 1 e: A consideration which elucidates or which suggests a particular (true or false) view of a subject. Hence, the aspect in which anything is viewed or judged. **10.** A window or other opening in a wall for the admission of light. **b.** *Gardening.* One of the glazed compartments (usually admitting of being opened) forming the roof or side of a greenhouse or the top of a frame. **11.** *Mech.* An aperture or clear space. **12.** *Painting.* Light or illuminated surface, as represented in a picture, or considered in regard to such representation; any portion of a picture represented as lighted up. **13.** *Law.* The light which falls on the windows of a house from the heavens, and which the owner claims to enjoy unobscured by obstructions erected by his neighbours. **14. a.** A flame or spark serving to ignite any combustible substance. **b.** Something used for igniting; e.g. a spill, taper, match.

When one is tempted to generalize on the meaning of the noun *light*, chiefly in modern times, and as used also in Middle English and sketchily in Old English, it must be remembered that the picture of any *Weltanschauung* that might seem to emerge from an inspection of the headings and subheadings of a dictionary entry is to be modified by the frequency with which each sense and subsense is used, and that frequency is not apparent from any dictionary entry. At the very beginning of the last century those working with Henry Bradley, including probably Bradley himself, had stacks of quotation slips on their desks, and from these stacks they selected a small number of quotations to print under the various senses and subsenses of the word *light*. In contemplating the meaning of *light* a general impression emerges: our current use of *light* is more anthropocentric, more involving our visual experience as human beings, than OE *leoht*, especially as used in verse.

'Light': 'the inward revelation of Christ in the soul'

Grein began by equating the word *leoht* with Latin *lux*, 'light'.[17] He did not choose *lumen* < **leuk-s-men*, a derivative of the same stem that gave *lux* and *leoht* in Latin and Old English.[18] A further derivative is *illuminatio*, which in the Oxford Psalm is equated with the Lord God. In

---

[17] See above, n. 13.
[18] For the etymology of these words, see Julius Pokorny, ed., *Indogermanisches etymologisches Wörterbuch*, 2 vols (Berne: Francke, 1959–1969), vol. I, p. 688, s.v. *leuk-*; A. Walde, ed., *Lateinisches etymologisches Wörterbuch*, 3rd edn, rev. J. B. Hofmann, 3 vols (Heidelberg: Carl Winter, 1938–1956), vol. I, pp. 823–24 s.v. *lūceō*, pp. 832–33 s.v. *lūmen*.

view of the fact that, as far as we know, all Old English verse was created and transmitted in monastic institutions, one might have expected poetic uses of OE *leoht* to be theocentric, comparable to the Quakers' use of *light* in sense 7. b. of the *NED* entry *Light*, 'Among Quakers, the inward revelation of Christ in the soul.' There are such occurrences, for example, (ignoring spelling differences):[19] *heofones leoht* 'light of heaven' (*Christ and Satan*, line 310, Vercelli *Homiletic Fragment*, line 44; and cf. *heofonleoht* 'heavenly light' at *Andreas*, line 974); *dryhtnes leoht* 'the Lord's light' (*Christ and Satan*, line 68, *Guthlac A*, line 583), *wuldres leoht* 'the light of glory' (*Christ and Satan*, lines 140, 251, 447, 555, 616, 648, *Andreas*, line 1611, *Fates of the Apostles*, line 61, *Guthlac*, line 8), *swegles leoht* 'light of the firmament' (*Christ and Satan*, line 28, *Guthlac A*, line 486, *Phoenix*, line 288, *Death of Edward*, line 28). To these may be added genitival phrases with *leoma* 'light', such as *wuldres leoma* (*Christ and Satan*, line 85), and *swegles leoma* (*Christ and Satan*, line 350, *Phoenix*, line 103), and the compound *heofonleoma* (*Andreas*, line 838).

'Light of light' in the Nicene Creed

Some of these uses are reminiscent of the wording of the Creed issued at the First Council of Nicaea in AD 325, the Nicene Creed. God is 'light of light': Φῶς ἐκ φωτός—'light from light', *lumen de lumine*, the Latin in books written, for example, in England in the Red Book of Derby of 1061 or shortly thereafter, and the Missals of the later Middle Ages.[20] All Anglo-Saxons were required to teach their children, in the vernacular if they knew no Latin, the Lord's Prayer and the Creed, but it was the Apostles' Creed, not the Nicene Creed, which is not found in English before the Benedictine Reform of the later tenth century.[21] There are

---

[19] All references to Old English poetic texts are (unless otherwise stated) to George Krapp and Elliott Van Kirk Dobbie, eds, *The Anglo-Saxon Poetic Records*, 6 vols (New York: Columbia University Press, 1931–1953): I, *The Junius Manuscript*, 1931; II, *The Vercelli Book*, 1932; III, *The Exeter Book*, 1936; IV, *Beowulf and Judith*, 1953; V, *The Paris Psalter and the Meters of Boethius*, 1932; VI, *The Anglo-Saxon Minor Poems*, 1942, except that *Beowulf* is quoted (unless otherwise stated) from R. D. Fulk, Robert E. Bjork, and John D. Niles, eds, *Klaeber's Beowulf and The Fight at Finnsburg* (Toronto: University of Toronto Press, 2008).

[20] For the texts see John N. D. Kelly, *Early Christian Creeds* (3rd edn; London: Longman, 1972), p. 215; Warren, trans., *The Sarum Missal in English*, p. 275; Legg, ed., *The Sarum Missal*, p. 211.

[21] For references to the Nicene Creed in late Old English: see Hans Sauer, ed., *Theodulfi Capitula in England*, Münchener Universitäts-Schriften, Texte und Untersuchungen zur Englischen Philologie, 8 (Munich: Wilhelm Fink, 1978), p.

therefore not many translations of the Nicene Creed into Old English containing the words *leoht of leohte*.[22]

Glossed versions exist of a hymn, probably to be ascribed to Ambrose, the first stanza of which has, in Old English, wordplay on *leoht*, and its verb *onleohtan*, the lemmata of which are *lux, lumen*, and its verb *illuminare*:[23]

> Eala, ðu beorhtnyss fæderlices wuldre
> of leohte leoht, forðbringende,
> leoht lucis & fons luminis,
> dæg dagena onleohtende.

[Oh, thou brightness of paternal glory, bringing forth light from light, light of light and source of light, lighting up the day of days.]

Frequently Old English glossing is element by element, here *on-leoht-ende* corresponds element by element to *in-lumin-ans*:

> O, splendor paterne gloriæ
> de luce lucem proferens,
> lux lucis & fons luminis,
> dies dierum inluminans.

333, and cf. p. 154; Felix Liebermann, ed., *Gesetze der Angelsachsen*, 3 vols (Halle: Max Niemeyer, 1898–1916), vol. I, p. 302 I Cn. 22; Dorothy Bethurum, ed., *The Homilies of Wulfstan* (Oxford: Clarendon Press, 1957), p. 157, VII, line 9, and p. 209, XC, line 170; Karl Jost, ed., *Die «Institutes of Polity, Civil and Ecclesiastical»*, Swiss Studies in English, 47 (Berne: Francke, 1959), p. 186 D. 22—cf. Roger Fowler, ed., *Wulfstan's Canons of Edgar*, EETS, o.s. 266 (London, 1972), pp. 28–30.

[22] A text from Bodleian MS Hatton 114 is in Max Förster, 'Die altenglischen Bekenntnisformeln', *Englische Studien*, 75 (1942), 159–69 (pp. 168–9). Similarly in Transitional English, *liht of lihte*, from Bodleian MS Junius 121 (with a facsimile of the Creed) in S. J. Crawford, 'The Worcester Marks and Glosses of the Old English Manuscripts in the Bodleian Together with the Worcester Version of the Nicene Creed', *Anglia*, 52 (1928), 1–25 (p. 5).

[23] Inge B. Milfull, ed., *The Hymns of the Anglo-Saxon Church*, Cambridge Studies in Anglo-Saxon England, 17 (Cambridge: University Press, 1996), p. 142; Helmut Gneuss, ed., *Hymnar und Hymnen im englischen Mittelalter*, Buchreihe der Anglia, 12 (Tübingen, Max Niemeyer, 1968), p. 281. Gneuss (p. 211) mentions that the first fifteen lines of the fifth Advent Lyric have been considered a parallel (cf. Jackson J. Campbell, ed., *The Advent Lyrics of the Exeter Book* (Princeton: Princeton University Press, 1959), pp. 55, 89 note on *god of gode*), but the parallel is not close.

Old English kennings based on 'light', and various tralatitious uses

In her account of Old English kennings, Hertha Marquardt discusses *heofonleoma*, *swegles leoma*, and *swegles leoht* as kennings.[24] Like French *ciel* < Latin *caelum*, or German *Himmel*, Old English does not distinguish 'heaven' from 'sky' conceptually; not that poetically this distinction is always clear in Modern English, as when near the end of the play within the play, *Midsummer Night's Dream*, V. i. 304–10, Pyramus makes his great suicide speech, with a use of *light*, the sense of which is far from clear to me, unless it means 'gift of life'; perhaps 'my soule is in the sky' is designed to exemplify Pyramus' or Bottom's verbal misplacing, that is, his lexicological inadvertency:[25]

> Come teares, confound: Out sword, and wound
> The pap of *Piramus*:
> I, that left pap, where heart doth hop;[26]
> Thus dye I, thus, thus, thus.
> Now am I dead, now am I fled, my soule is in the sky,
> Tongue lose thy light, Moone take thy flight,
> Now dye, dye, dye, dye, dye.

If the definition of the term kenning is of a phrase or compound which involves some leap of the imagination, such as calling the sun *heofoncondel*, that is, 'the candle or taper of heaven', I fail to see how the factual descriptions *swegles leoma* and *swegles leoht*, 'light of the sky', can be regarded as kennings for the sun.[27] To justify her interpretation Marquardt suggests that here we can readily equate *leoht* and *swegl* — as

---

[24] Hertha Marquardt, *Die altenglischen Kenningar*, Schriften der Königsberger Gelehrten Gesellschaft, 14th year, Geisteswissenschaftliche Klasse, 3 (Halle: Max Niemeyer, 1938), pp. 184–86 = 82–84 of separate.

[25] Charlton Hinman, ed., *The Norton Facsimile: The First Folio of Shakespeare* (New York: W. W. Norton & Company, 1968), p. 179.

[26] In this line I is a spelling for ay; and pap probably rhymes with hop in Bottom's pronunciation.

[27] Hendrik van der Merwe Scholtz, *The Kenning in Anglo-Saxon and Old Norse Poetry* (Utrecht doctoral dissertation, 1927), pp. 68–69 similarly regards these phrases as kennings, and even considers *leoht*, at Beowulf, line 569 (Fulk, Bjork, and Niles, eds, Klaeber's Beowulf, p. 21), as a kenning. That this is contrary to what is usually understood by that term, well defined by Richard M. Meyer, *Die altgermanische Poesie nach ihren formelhaften Elementen beschrieben* (Berlin: Wilhelm Hertz, 1889), p. 157, *Umschreibung vermittelst variirter Appellativa*, that is, 'circumlocution by means of appellatives in variation with what is being designated'.

also more persuasively *wuldor*, 'glory' — with 'heavenly majesty' and even with 'God' himself. She discusses the matter further, assuring the reader that the Anglo-Saxons did not imagine heaven spatially; rather that the heavenly glory was expressed by circumlocution, and that therefore *leoht* is not to be understood literally.[28] The certainty is admirable with which Anglo-Saxonists can tell when the poets have abandoned what looks like a reality, when they have gone over to a tralatitious, entirely figurative meaning. At this point Marquardt treats not only of *dryhtnes leoht*, *swegles leoht*, and *wuldres leoht*; but also of *Godes leoht* (*Guthlac B*, line 1369, Krapp and Dobbie, III, 88), *leoht Godes* (*Beowulf*, line 2469, Fulk, Bjork, and Niles, eds, *Klaeber's Beowulf*, pp. 85, 247 note on the line).[29] The two uses are similar in that each is the account of a man's dying, Guthlac's and Hrethel's, the soul's entry into heaven or hope of entry into heaven conceived of as spatial, as it seems. First Guthlac's death (lines 1366b–1370a):

>   Nu se eorðan dæl,
> banhus abrocen,   burgum in innan
> wunað wælræste,   ond se wuldres dæl
> of licfæte   in leoht Godes
> sigorlean sohte.

[Now the terrestrial part, broken in body, lies on (his) death-bed within the dwellings, and the part of glory away from the bodily enclosure has sought glorious reward in God's light.]

This translation is brutally literal, and virtually meaningless when *in leoht Godes* is rendered 'in God's light'. I think what it means is 'in life with God', but life in death is a notorious oxymoron. The word *leoht* is to be understood spatially here, in God's radiant habitation, where the radiance shines forth, not from one of the bodies celestial which are visible from our earth, but from the glory of God in mandorla, as he was depicted in illuminations from the early Middle Ages onwards. Freed from the

---

[28] Marquardt, *Die altenglischen Kenningar*, pp. 305–06 = pp. 203–04.
[29] Friedrich Klaeber, ed., *Beowulf and The Fight at Finnsburg* (3rd edn; Boston, Massachusetts, later issues Lexington, Massachusetts, 1950), p. cx, note 3, comments on 'The somewhat strange expression applied to Hrēðel's death, and points to what he regards as parallels in *Genesis A*, but does not mention Guthlac's death. Hugh Magennis, 'Imagery of Light in Old English Poetry: Traditions and Appropriations', *Anglia*, 125 (2007), 181–204 (pp. 200–2) discusses some verse passages in which *leoht* stands for 'life'.

enclosing vessel of flesh and bone Guthlac's soul sought a new existence, a new life, in the light of God.

Now to the death of Hrethel, Beowulf's grandfather on the maternal side, and the earliest of the Geats, Beowulf's family, as recorded in the poem (lines 2468–71):

> He ða mid þære sorhge　þe him sio sar belamp,
> gumdream ofgeaf:　Godes leoht geceas;
> eaferum læfde,　swa deð eadig mon,
> lond ond leodbyrig,　þa he of life gewat.

[He then with the grief, which that sorrow assigned to him, gave up the joy of mankind: he chose God's light; he left the members of his court, as does one blessed, his lands and the nation's capital, when he departed this life.]

Hrethel's sorrow was caused by his sons: the accidental killing of Herebeald by his younger brother Hæthcyn. Klaeber (p. xxxviii) assigns dates to Hrethel, AD 445–503, but (at p. xxxi) he admits that such dates 'at best could represent approximate dates only'. For a man dying in pre-Christian times, what does *eadig mon* mean, 'one blessed'? The pagan Hrethel, unlike Guthlac, was no saint; but the poet was a Christian; and that is why, when he speaks of God as he does not infrequently, he is likely to have meant the God venerated by him as a Christian, and when he called a dying pagan *eadig*, 'blessed', he is likely to have thought of him as a good man who deserved a place in the bliss of heaven. In his imagination, as in the illuminations he may have seen, God was encircled in a mandorla, and in his imagination the blessed Hrethel, having left behind such joys as his court, his capital, and his country could give him, chose a new life in the light of God. Unlike a modern scholar, who assigns approximate dates to figures in the poem to enable the reader to put them into some kind of historical order, the poet of *Beowulf* is likely to have had little sense of a history with dates, and no sense of the nature of anachronism.

OE leoht in Genesis B

A much discussed use of *leoht* in Old English poetry is taken from *Genesis B*, lines 306b–312, the Fall of the Angels:[30]

---

[30] For this passage the following editions (with glossary or relevant notes or both) have been consulted: Eduard Sievers, ed., *Der Heliand und die angelsächsische Genesis* (Halle: Lippert'sche Buchhandlung, Max Niemeyer, 1875); Friedrich Klaeber, ed., *The Later Genesis*, Englische Textbibliothek, new edn (Heidelberg: Carl Winter, 1931); Benno J. Timmer, ed., *The Later Genesis* (Oxford: Scrivener Press, 1948);

Feollan þa ufon of heofnum
þurhlonge [31] swa þreo niht and dagas
þa englas of heofnum on helle, and heo ealle forsceop
Drihten to deoflum. Forþon heo his dæd ond word
noldon weorðian, forþon he heo on wyrse leoht
under eorðan neoðan ællmihtig God
sette sigelease on þa sweartan helle.

[Then they fell thus from heaven above for very long, three nights and days, those angels from heaven into hell, and the Lord deformed them all into devils. Because they would not honour his action and words, therefore he established them in worse light inglorious in defeat beneath the earth in that black hell.]

Sievers established that *Genesis*, lines 235–851, differed from the rest of the poem, and was derived from Old Saxon, an inference based on the closeness of lines of this part of *Genesis* to the Old Saxon *Heliand*, in lexis and metre. He was proved right by the discovery, about twenty years later, of Old Saxon *Genesis* fragments in the Vatican Library.[32] Sievers was careful in his identification of Old Saxonisms in the Anglo-Saxon *Genesis B*. He never said that *on wyrse leoht* was an Old Saxonism. In a footnote to an earlier occurrence in the poem, line 258, *on þam leohte*, likewise line 851, and similarly line 508, Satan's false account how God, *on his leohte*, praised Adam's *dæd and word*. Sievers renders *on þam leohte* as *in dieser Welt*, 'in this world', and compares Paris Psalter, Psalm 55, last line, *on lifgendra leohte*.[33] He compares with such Old English uses the frequent Old Saxon uses of the cognate in the transferred senses, 'life, world,

---

Dorothy Whitelock, ed., *Sweet's Anglo-Saxon Reader*, rev. edn (Oxford: Clarendon Press, 1967), pp. 130, 349; A. N. Doane, ed., *The Saxon Genesis* (Madison, Wisconsin: University of Wisconsin Press, 1991).

[31] For the interpretation that *þurhlonge* is calqued on Latin *perlonge*, see Ernst A. Kock, *Plain Points and Puzzles: 60 Notes on Old English Poetry*, Lunds Universitets Årsskrift, n.s., class 1, vol. 17, nr. 7. (Lund: C. W. K. Gleerup, 1922), pp. 12–13.

[32] See Karl Zangemeister and Wilhelm Braune, eds, 'Bruchstücke der altsächsischen Bibeldichtung aus der Bibliotheca Palatina', *Neue Heidelberger Jahrbücher*, 4 (1894), 205–94 for the Old Saxon texts. Unique in Germanic studies, this confirmation as fact of a Neo-Philological inference, resulted in Sievers being looked upon by many scholars as virtually infallible for some thirty years (after which he advocated *Schallanalyse*, the rhythmical analysis of verse and rhythmical prose in the many languages familiar to him).

[33] Krapp and Dobbie 1931–1953: V, 8, Psalm 55 verse 10 line 8. The occurrence had been listed by Grein, *Sprachschatz*, vol. II, p. 177, s.v.; see above, n. 13.

earth', especially with reference to God and heaven. There are very many occurrences in *Heliand,* and five in the Vatican *Genesis,* for example, line 76b, *so lango so thu thit liaht uuaros,* 'as long as thou didst inhabit this light'.[34] A significant use in *Heliand* of the sense 'light' is line 487, part of the (extended) Song of Simeon: *Thu bist lioht mikil* || *allun elithiodun* | *thea er thes Alouualdon* || *craft ne antkendun,* 'Thou art a great light to all Gentiles who had not cognized the might of the Omnipotent'[35]; based on Luke 2. 32, *lumen ad reuelationem gentium et gloriam plebis tuae Israhel,*[36] rendered in the Rhemes Version, 'A light to the reuelation of the Gentils, and the glorie of thy people Israel.'[37] In Psalm 55. 13 the Psalmist refers to this life here: the Vulgate has: *ut placeam coram Deo in lumine viventium,*[38] rendered in the Doway Version 'that I may please before God, in the light of the liuing.'[39] It seems that, via the Vulgate, we are dealing with a Hebraism.[40] 'Light', *leoht* in Old English, *lumen* and *illuminatio* in the Latin of the psalter (in which the monastic poets of Anglo-Saxon England were at home) cannot be the property of any one national tradition. It occurs with some frequency in the Old Saxon poetic records, and occurs several times in the Old English *Genesis B*. That does not mean that this particular sense in Old English came from Old Saxon; yet that is what most twentieth-century editors of and commentators on

---

[34] Doane, ed., *The Saxon Genesis,* p. 239. Cf. Edward H. Sehrt, ed., *Vollständiges Wörterbuch zum Heliand und zur altsächsischen Genesis,* Hesperia, XIV (1925, repr. Göttingen: Vandenhoeck & Ruprecht, 1966), pp. 342–43.

[35] Eduard Sievers, ed., *Heliand* (Halle: Buchhandlung des Waisenhauses, 1878; reissued with appendix by Edward Schröder, 1935), p. 37.

[36] John Wordsworth, Henry Julian White et al., eds, *Nouum Testamentum Domini Nostri Iesu Christi Latine,* 3 vols (Oxford: Clarendon Press, 1889–1954), vol. I, p. 319.

[37] *The New Testament of Iesus Christ, Translated Faithfully into English out of the authentical Latin ... in the English College of Rhemes* (Rhemes: by Iohn Fogny, 1582), p. 141.

[38] *Biblia Sacra iuxta Latinam Vulgatam versionem ad codicum fidem,* 18 vols (Rome: Typis Polyglottis Vaticanis, 1926–1995), vol. X *Liber Psalmorum,* p. 142.

[39] *Holie Bible* (Doway, 1609, 1610), vol. II, p. 108, Ps. 55:13.

[40] It has been shown (by Alan S. C. Ross, 'OE. "leoht" 'world"', *Notes and Queries,* 220 (1975), 196) that this metaphorical use of 'light' is very widespread in the languages of the world. With the occurrences in *Genesis B,* lines 310 and 851 listed by Ross, that in line 258 should be compared; and his denial that this use was not calqued on Latin or Greek does not apply to biblical Latin and Greek.

the Old English poem aver.[41] The Old Saxon poets used the word in this sense quite frequently, the Old English poets used the word in this sense relatively rarely.

Chaucer's Parson quoting the Sermon on the Mount may invite exegesis

Many religious thoughts come to one when speaking of light, and I shall conclude with one such thought on the Sermon on the Mount (Matthew 5. 14–16), quoted by Chaucer's Parson (Fragment X (I) 1036–37):[42]

> A citee may nat been hyd that is set on a montayne, ne men lighte nat a lanterne and put it vnder a busshel, but men sette it on a candle-stikke to yeue light to the men in the hous. Right so shal youre light lighten bifore men, that they may seen youre goode werkes, and glorifie youre Fader that is in heuene.

Commentaries on the Sermon on the Mount are many, and I quote not only a little of what the Rhemes Version abstracted from the Fathers, but also how the Puritan William Fulke could not here find much to controvert in the popish teaching of the Rhemes Seminary. Chaucer's Parson quotes the Gospel to teach Penitence towards which 'youre goode werkes' are directed. The Fathers of the Church have a different exegetical purpose, and begin with 'You are the light of the world', the sentence preceding the Parson's quotation. The Rhemes annotation of Matthew 5:11 under the marginal heading 'The Church visible', states (quoting St Augustine):[43]

> The light of the world, and citie on a mountaine, and candle vpon a candlesticke, [...] signifieth the Clergie, and the whole Church, so built vpon Christ the mountaine, that it must needes be visible, and can not be hid nor vnknowen [...] And therfore, the Churche being a candle not vnder a bushel, but shining al in the house (that is) in the world.

'Your light' towards the end of the Parson's quotation is annotated in the Rhemes Version on verse 17, 'The good life of the Clergie edifieth much,

---

[41] See, for example, the unidirectional comments by Timmer, ed., *The Later Genesis*, p. 34, where the word at line 310, under his Group C, as a word 'used in a special sense in our poem, but which occur in Old Saxon'; and Doane, ed., *The Saxon Genesis*, p. 264, as if the word existed uniquely in this poem in an 'OS sense, "world."'

[42] F. J. Furnivall, ed., *A Six-Text Print of Chaucer's Canterbury Tales*, part VIII, Chaucer Society, 1st series, XLIX (London, 1877), p. 679 (Ellesmere MS).

[43] *New Testament* (Rhemes, 1582), p. 14.

and is Gods great honour: whereas the contrarie dishonoureth him.' Fulke elaborates, under his marginal heading 'Visibilitie of the Church', that Augustine '[a]lso [...] acknowledgeth, that the Church may be so secret that the members knowe not one another [...] And the Catholike Church which is the whole mysticall bodie of Christ, being an article of faith, is alwayes inuisible.'[44] By 'the Catholike Church' Fulke of course means the Church Universal, that whole spiritual entity hallowed in the Church of England, together with all Christianity untainted by Popish error. Visibility and invisibility, the Clergy, that is the light, not hid under a bushel, but raised on a candlestick, that signifies, if we are to believe St Hilary as quoted in Thomas Aquinas' *Catena Aurea*, 'the lamp, i. e. Christ Himself, set on its stand when He was suspended on the Cross in his passion'.[45]

Chaucer's Reeve invites no holy thoughts

From the sanctity of anagogical exegesis, from debate made fierce by piety, I move to nearer the beginning of the Canterbury Tales, where piety enters only as an occasional curse: the real world of *The Reeve's Tale*, for I am not among its mystical explicators, though 'bushel', in this Tale 'half a bushel', is significantly present as it is in the Sermon on the Mount. The Fathers of the Church explicating the light in Matthew 5 make much of moonshine, which I have not quoted — see the *Catena Aurea*. Towards the end of *The Reeve's Tale* a little chink admits enough moonlight to illumine the bald pate of Symkyn, the miller of Trumpington. His wife had shouted that he should help her, for John, one of the two Fellows of a Cambridge College — later incorporated into Trinity — was lying on her belly; or was it Symkyn thrown down on her head by Aleyn, the other Cambridge Fellow, in their scuffle? In the dark, who can tell? (I (A) 4288–310):[46]

---

[44] *The Text of the New Testament of Iesus Christ, Translated Out of the vulgar Latine by the Papists of the traiterous Seminarie at RHEMES ... with A Confutation of all such Arguments, Glosses, and Annotations, As Conteine Manifest impietie, of heresie and slander, against the Catholike Church of GOD...* by William Fulke (London: by Christopher Barker, 1589), 9$^{\text{ro–vo}}$.

[45] John Henry Newman and Mark Pattison, trans., *Catena Aurea ... collected by S. Thomas Aquinas*, 4 vols in 8 parts (Oxford: John Henry Parker, 1841–1845), pt. I/1, pp. 162–65.

[46] F. J. Furnivall, ed., *A Six-Text Print of ... Canterbury Tales*, part II, Chaucer Society, 1st series, XIV (London, 1870), p. 123 (Hengwrt MS).

'Awake Symond / the feend is on me falle,
Myn herte is broken, / help! I nam but ded:
Ther lyth oon vp on my wombe / and vp myn hed.
Help Symkyn! / for the false clerkes fighte.'
¶ This Iohn sterte vp / as faste as euere he myghte,
And graspeth by the walles / to and fro
To fynde a staf. / And she sterte vp also,
And knew the estres / bet than dide this Iohn,
And by the wal / a staf she foond anon,
And saugh / a litel shymeryng of a light,
For at an hole / in shoon the moone bright,
And by that light she saugh hem bothe two;
But sikerly / she nyste who was who.
But as she saugh / a whit thyng in hir Iye;
And whan she gan / this white thyng espye
She wende the clerk hadde wered a voluper,
And with the staf / she drow ay ner and ner,
And wende han hit / this Aleyn atte fulle,
And smoot the millere / on the piled skulle
That doun he gooth, / and cryde, 'Harrow, I dye!'
Thise clerkes bette hym wel / and lete hym lye,
And greithen hem, / and tooke hir hors anon,
And eek hir mele, / and on hir wey they gon;
And at the mille / yet they toke hir cake
Of half a busshel flour / ful wel ybake.

['Simon, wake up! The devil has fallen on me, my heart is broken. Help! I am almost dead. There is somebody lying on my belly and on my head. Symkyn, help! for these wicked Fellows are fighting.' Now John gets up as quickly as ever he could, and feels his way along the wall hither and yon to find a stick. And she starts up too, and she knew the inside of the room better than John did; and at once she found a stick by the wall; and she saw just a slight reflection of some light, for the moon shone in brightly through a chink, and by that light she saw both of them; but truly she did not know which was which. But she saw in her eye a white thing, and when she first looked at this white thing she thought the Fellow was wearing a nightcap, and she got ever closer and closer with the stick, and hit the miller on his bald pate, so that he goes down, shouting, 'Help! I am dying.' These Fellows beat him up soundly, and left him lying, and got themselves dressed, and quickly took their horse, and also their flour, and on their way they go. And at the mill they further took their cake, very well baked of half a bushel of flour.]

If only, on the first day of Creation, God the All-knowing had foreseen the fracas at the mill in Trumpington, he might have divided day from night, the light from the darkness, less severely. This is one of the great 'ifs' of literary history. It would have diminished the literature of mistaken identity, and thus it would have spelt the end of farce before that genre had even begun. It would have left unknotted many comedies in which darkness could no longer enshroud the heroine, wronged only to be righted in full light of day, and then All's Well that Ends Well. And in tragedy many characters would have run their race differently; among them in *The Reeve's Tale* poor, forsaken Malyne, she with that sluttish name, as she bids farewell to Aleyn, 'God thee saue and kepe!' (line 4247). Her words, spoken in sorrow, provide a good, sadly bright note on which to end.

THE RELATIONSHIP OF LIGHT AND COLOUR IN MEDIEVAL THOUGHT AND IMAGINATION

Michael J. Huxtable

Introduction

The primary point of access for medieval thought concerning visual experience was theology: an overarching set of beliefs concerning the divine significance of light (*lux*) in accordance with the creation of the world at God's utterance presented in the first chapter of Genesis. To use Katherine Tachau's memorable phrase, visuality was 'the nexus of natural philosophy and epistemology, all ultimately at the service of theology'.[1] Whilst there is always room to question what exactly constituted 'visuality' for medieval thinkers and writers, it is clear that the role and nature of light was of paramount importance for Patristic writers onwards. Consequently, the significance of light left to colour, considered as a phenomenon separate from light, a secondary role and importance. Colour was over-shadowed, or over-exposed, by the more emphatically divine visual progenitor. John Gage, in his seminal study *Colour and Culture: Practice and Meaning from Antiquity to Abstraction* (1993), argues for this state of affairs emphatically:

> The relationship of light to colour was a matter of some debate in these centuries but there was a general agreement that colour was at best a secondary attribute of light, its most material aspect, accident rather than substance.[2]

Gage concludes:

> Colour was related to *lumen* and not *lux* and thus was at two removes from the highest form of light.[3]

---

[1] Katherine Tachau, *Vision and Certitude in the Age of Ockham: Optics, Epistemology and the Foundations of Semantics 1250-1345* (Leiden: E.J. Brill, 1988), p. xvi.

[2] John Gage, *Colour and Culture: Practice and Meaning from Antiquity to Abstraction* (London: Thames and Hudson, 1993), p. 70.

[3] Gage, p. 70.

Gage's position recalls the definitions of *lux* and *lumen* given by Isidore of Seville in the *Etymologies*: 'Light (*lux*) is the substance itself, while illumination (*lumen*) is so called because it emanates from light (*a luce manare*), that is, it is the brightness (or whiteness) of light (*lux*) - but writers confuse the two.'[4] The following article re-approaches the relationship of light and colour in later medieval thought, imagination and culture, to show how the conceptual significance of colour changed from the twelfth century in line with the increasing integration of Aristotelian intromissive perception into scholastic thought. In broad terms, the classical influence on medieval thought on visual perception and the nature of light can be summarized as an early 'extramissive' mode of discourse derived from Plato's *Timaeus*, and a later 'intromissive' discourse from Aristotle's *De anima*. Cultural and artistic ramifications of this paradigm shift can be argued to have included, amongst other phenomena, the rise of systematic heraldry. The relationship of colour to light based fundamentally upon its proximity to *lux*, a formula perhaps receiving its most extreme consequences at the hands of Bernard of Clairvaux,[5] gradually became replaced by one in which colours might be divinely or otherwise orientated, and their use and abuse adjudged independently according to a hierarchy of colour values.

Lux, lumen and divine significance

The difference between *lux* and *lumen* was a fundamental topic of theological discussion by early Christian writers, primarily in commentaries on the Hexameron (six days of creation) of Genesis 1–2. 2. In the first of the Genesis accounts light (*lux*) was created from the first utterance of God ('*Dixitque Deus fiat lux et facta est lux*')[6] and receives the

[4] *The Etymologies of Isidore of Seville*, trans. Stephen A. Barney, W. J. Lewis, J. A. Beach et al. (Cambridge: Cambridge University Press, 2006), XIII.x.14 (p. 274).

[5] Bernard of Clairvaux (1090–1153) regarded coloured phenomena as indicative of sinfulness. He memorably declared 'Caecitas colorum!' ('We are blinded by colours!') in various sermons. See also Michel Pastoureau, *L'Eglise et la couleur, des origines à la Réforme*, Bibliothèque de l'École des chartes 147 (1989), 203–230 (pp. 207–8); and Herman Pleij, *Colors Demonic and Divine: Shades of Meaning in the Middle Ages and After*, trans. Diane Webb (New York: Columbia UP, 2002, 2004) 3; Cistercian attitudes towards colour and ornamentation are discussed in G. Duby, *Saint Bernard et l'art cistercien*, Arts et Métiers Graphiques (Paris: Flammarion, 1979).

[6] 'And God said Let there be Light: and there was Light.' Gen.1.3. The Vulgate term *lux* translates the Hebrew word אור ('ôr), and the Greek φῶς (phōs)—used in John's Gospel to define the divine light (Christ) and used in the Septuagint to

first benediction. Thus it was held by early Christian writers including Basil the Great (c.329–379) that light was an intrinsic property of matter and a direct emanation of God.⁷ This mode of thinking was significantly developed by St Augustine of Hippo (354–430) who argued that God's primary light (*lux*) was distinguishable from that which derived secondarily from heavenly bodies (*luminaria*). Key to this understanding is the application of Platonic 'ideas' or 'forms' (and human access to them) to the account of divine creativity as a process of 'illumination'. Plato's *Timaeus* had included the idea of an elementally construed corporeal light existing as both an external and internal fiery substance. According to this theory, vision occurs when an internally derived ray of light is emitted from the eye and coalesces with the external fire of the atmosphere; the interruption of the coalescence by the presence of an object sending tremors back along the ray thereby communicating the form of an object to the soul. The extramissive model presents therefore 'seeing' as a species of touch, achieved via internal and external 'light' qua the element of fire. The key source is paragraph 45b of the *Timaeus*:

> The eyes were the first organs to be fashioned by the gods, to conduct light. [...] They contrived that such fire as was not for burning but for providing a gentle light should become a body, proper to each day. Now the pure fire inside us, cousin to that fire, they made to flow through the eyes. [...] Now whenever daylight surrounds the visual stream, like makes contact with like and coalesces with it to make up a single homogeneous body aligned with the direction of the eyes. This happens whenever the internal fire strikes and presses against an external object it has connected with: [...] it transmits the motions of whatever it comes in contact with as well as whatever comes in

---

translate אוֹר ('ôr). The Greek term could be used for fire-light but came to be used primarily figuratively, especially of light conceived as a ray. Lumen, by contrast, translated λύχνος (luchnos), and was a term employed for light derived from an indirect source such as a lamp or candle. The semantic difference can be seen in verses such as John 5. 35: 'John (the Baptist) was a burning and shining light (luchnos) and you were willing, for a season, to rejoice in his light (phōs).' The verse shows the complexity of the matter for theological exegesis: John is presented as both a metaphorical light source and a derivative light. His derivative 'light', in which followers are exhorted to rejoice, is phōs, suggesting a non-derivative light. However, he as a 'source' of light is termed luchnos—possibly underlining John's secondary status in relation to the divine light.

7 Saint Basil, Bishop of Caesarea, *The Syriac Version of the Hexaemeron by Basil of Caesarea*, trans. and ed. Robert W. Thomson, Corpus Scriptorum Christianorum Orientalium, vols 550–51; Scriptores Syri Tomus 222–23 (Lovanii: Peeters, 1995) *Homily* 6, pp. 85–112; trans. vol. 551, pp. 71–93.

contact with it, to and through the body until they reach the soul. This brings about the sensation we call 'seeing'. At night, however, the kindred fire has departed and so the visual stream is cut off.[8]

The notion of a 'visual stream' and an internal fire 'striking' and 'pressing against' the external object offers a fundamentally subject-orientated, tactile model of visual perception. On the extramissive model, light-orientated vision could be employed to make sense of divine creativity and the connection of the supernatural realm to natural phenomena. Integral to Augustine's thinking was a comparable extramissive model of visual perception, and the view that both an 'external' and 'internal' light are required for vision:

> iactus enim radiorum ex oculis nostris cuiusdam quidem lucis est iactus et contrahi potest, cum aerem, qui est oculis nostris proximus, intuemur, et emitti, cum ad eandem rectitudinem quae sunt longe posita adtendimus. Nec sane, cum contrahitur, omnino cernere, quae longe sunt, desinit, sed certe obscurius, quam cum in ea obtutus emittitur. Sed tamen ea lux, quae in sensu uidentis est, tam exigua docetur, ut, nisi adiuuetur extraria luce, nihil uidere possimus.[9]

> [surely the emission of rays from our eyes is an emission of a certain light. And it can be gathered that this [light] is emitted, since when we look into the air adjacent to our eyes we observe, along the same line, things situated far away. Nor does this light sensibly fail, since it is judged to discern fully objects that are at a distance, though surely more obscurely than if the power of sight should [itself] be sent to them. Nevertheless, this light that is in vision is shown to be so scanty that unless it is assisted by an exterior light, we cannot see anything.][10]

Under Timaean influence, therefore, Augustine understood visual perception to require two kinds of light: the interior and exterior. Moreover, for the pre-eminence of the divine light to be asserted, the light emitted from the human eye had to be of an inferior kind to that originating from the deity. Thus Augustine (a former Manichaean[11]) fused the Christian

---

[8] Plato, *Timaeus*, trans. Donald J. Zeyl (Cambridge/Indianapolis: Hackett, 2000), p. 33, para. 45b–d.

[9] Augustine, *De Genesi ad litteram libri duodecim*, ed. Joseph Zycha, CSEL, vol. 28, part 1 (1894), p. 23.

[10] Translation from Lindberg, p. 90.

[11] Augustine converted to Christianity in 387 (see *The Confessions* books 7 and 8 for his autobiographical account of this). It has been argued that Augustine's

theology of the *logos* with that of the principal *ego eimi* claim of 'the light of the world' and paralleled this with the *Timaean* position of sight having been created in order to provide mankind with understanding.[12] Discussing Genesis 1. 2 ('But the earth was invisible and without form') in *De Genesi contra Manichaeos*, Augustine makes a clear distinction between the differing kinds of light in order to refute the argument made by the Manichaeans that before God created light he must have existed in darkness:

> Vere ipsi sunt in tenebris ignorantiae et ideo non intellegunt lucem, in qua deus erat, antequam faceret Istam lucem. Non enim norunt isti lucem nisi quam carneis oculis vident, et ideo istum solem quem pariter non solum cum bestiis maioribus, sed etiam cum muscis et vermiculis cernimus, illi sic colunt, ut particulam dicant esse lucis illius in qua habitat deus. Sed nos intellegamus aliam esse (p. 73) lucem in qua deus habitat, unde est illud lumen, de quo in evangelio dicitur: *erat lumen verum quod illuminat omnem hominem venientem in hunc mundum.* Nam solis istius lumen non illuminat omnem hominem, sed corpus hominis et mortales oculos, in quibus nos vincunt aquilarum oculi, qui solem istum multo melius quam nos dicuntur aspicere. Illud autem lumen non irrationabilium avium oculos pascit, sed pura corda eorum, qui deo credunt et ab amore visibilium rerum et temporalium se ad eius praecepta implenda convertunt; quod omnes homines possunt si velint, quia illud lumen *omnem hominem illuminat venientem in hunc mundum.*[13]

[They themselves [the Manichaeans] are truly in the darkness of ignorance, and for that reason they do not understand the light in which God was

---

Manichaean past influenced his Christian theology in numerous ways, perhaps the most obvious concerning the pre-eminence of light in his thinking about the divine. The Manichaeans, who originated in Persia (and were in turn influenced by Zoroastrianism), worshipped a pantheon of divine spirits of light and perpetually struggled with spiritual forces of darkness. However, this line of thought—attributing Augustine's emphasis on divine light to his Manichaean background—runs the risk of underplaying the distinct role of light as used theologically in the Gospels, and its influence upon readers and interpreters. See Paul Mirecki and Jason BeDuhn, eds, *The Light and the Darkness: Studies in Manichaeism and its World* (Leiden: Brill, 2001).

[12] Cf. John 1. 1: the divine logos or second person of the Trinity pre-existed the world, but in terms of Greek philosophical tradition would also imply the meaning or final telos of the world.

[13] Augustine, *De Genesi contra Manichaeos*, III. 6, ed. Dorothea Weber, Sancti Augustini Opera, CSEL, 91 (Wien: Verlag der Österreichischen Akademie der Wissenschaften, 1998), pp. 72–73.

before he made this light. For they know only the light they see with the eyes of the flesh. And therefore they worship this sun which we see, not only along with the larger animals, but even with flies and worms, and they say that this sun is a particle of that light in which God dwells. But let us understand that there is a different light in which God dwells. From it there comes that light of which we read in the Gospel [...] (p. 54) For the light of this sun does not enlighten all of man, but the body of man and his mortal eyes, in which we are surpassed by the eyes of eagles which are said to gaze upon this sun much better than we. But that other light feeds, not the eyes of irrational birds, but the pure hearts of those who believe God and turn themselves from the love of visible and temporal things to the fulfilment of his commands. If they wish to, all men can do this, because that light enlightens every man coming into this world.][14]

Augustine's differentiation between the light of mortal '*carneis oculis*', which human beings shared with other animals (albeit in weaker form compared to eagles), and the '*lucem in qua deus habitat*' entailed two kinds of light:

> the light that is perceived by the eyes is one thing; the light which acts through the eyes so that sensation might occur is something else.[15]

In other words, first there had to be a difference in kinds between the external light provided by the sun and other light sources and that of a living subject's 'eyebeam'. However, to further clarify his understanding of the inferior nature of fleshly vision Augustine went on to think in terms of three kinds of light. In the unfinished work *De Genesi ad litteram*, Augustine considered how humankind and animals share fleshly eyes but not the faculty of rationality, which was the 'true' essence and potential for illumination and visual perception. He distinguished, therefore, between the inner and outer lights needed for extramissive seeing and another type required by contrast because, 'Even the souls of other animals do not lack such light.'[16] The other light 'can be understood in creatures, that by which they reason. To this is opposed as darkness the irrationality, such as is found in the souls of other animals'.[17]

---

[14] *Saint Augustine on Genesis: Two Books on Genesis – 'Against the Manichees' and 'On the Literal Interpretation of Genesis: An Unfinished Book'*, trans. Roland J. Teske, The Fathers of the Church, 84 (Washington, D.C.: Catholic University of America Press, 1991), pp. 53–54.

[15] Augustine, *De Genesi ad litteram*, trans. Teske, p. 159.

[16] Augustine, *De Genesi ad litteram*, trans. Teske, p. 157.

[17] Augustine, *De Genesi ad litteram*, trans. Teske, p. 160.

Thus Augustine cited the human ability to reason as an experience of the divine provision of a kind of light, that which ultimately distinguished humanity, made in the image of God (Genesis 1. 26–7), from beasts.

Augustine's religiously orientated triad of kinds of light was clearly intended to be understood simultaneously as a metaphor and as a spiritual reality, and its influence permeates medieval and even later theological and philosophical writings.[18] However, the extramissive Platonist aspect of the position in relation to celestial perception—the light of reason grounded in divine Truth—achieved its most notable expression in Pseudo-Dionysius' fifth century *De coelesti hierarchia* (or *The Celestial Hierarchy*) in which the author charted the emanation and path of divine light as it encompassed the celestial order, illuminating different levels of supernatural beings. Pseudo-Dionysius paralleled the celestial order on earth with the natural hierarchy of men and the world of beasts, and the ecclesiastical hierarchy of church offices (the latter treated separately in *The Ecclesiastical Hierarchy*). *The Celestial Hierarchy* opens in quotation of the epistle of St James' description of God as 'the Father of Lights' (James 1. 17). This formulation is taken further so that:

> every divine procession of radiance from the Father, while constantly bounteously flowing to us, fills us anew as though with a unifying power, by recalling us to things above, and leading us to the unity of the Shepherding Father and to the Divine One.[19]

The extramissive procession of radiance—linking man to God—served as a basis for the portrayal of an exemplary hierarchy operating within heaven itself. The structure is triadic: nine types of beings sorted into

---

[18] A theologically sophisticated fourteenth-century use is that of Thomas à Kempis' prayer for mental illumination in the *Imitatio Christi*, Book III, Capitulum 27: 'Claryfie me, [gode Ihesu,] with þi clerenes of euerlastynge light and bringe oute of þe habitacle of my herte alle maner of derkenes. Sende oute þi light and þi trouþe, þat þei mowe shyne vppon þe yerthe, for I am ydel [or vayne] yethe and voyde, till þou illumine me [...]' (*The Imitation of Christ: The First English Translation of the 'Imitatio Christi'*, ed., B. J. H. Biggs, EETS, o.s. 309 (Oxford: Oxford University Press, 1997), p. 96). The text of this ME edition is composed from three translations dating from the fifteenth and early sixteenth centuries (see Preface, p. vii).

[19] Pseudo-Dionysius, *The Celestial Hierarchy*, trans. Colm Luibheid, in *Pseudo-Dionysius: The Complete Works*, trans. Colm Luibheid, forward, notes, introductions by Paul Rorem, Rene Roques, Jaroslav Pelikan et al., Classics of Western Spirituality (New York: Paulist Press, 1987), 121A, pp. 145–191 (p. 145).

three groups of three; thrones, cherubim and seraphim at the highest level; angels, archangels and principalities at the lowest:

> The word of God has provided nine explanatory designations for the heavenly beings, and my own sacred initiator has divided these into three threefold groups. According to him, the first group is forever around God and is said to be permanently united with him ahead of any of the others, and with no intermediary. Here, then, are the most holy 'thrones' and the orders said to possess many eyes and many wings, called in Hebrew the 'cherubim' and 'seraphim'. Following the tradition of scripture, he says that they are found immediately around God and in a proximity enjoyed by no other. This threefold group says my famous teacher, forms a single hierarchy which is truly first and whose members are of equal status. No other is more like the divine or receives more directly the first enlightenment from the Deity.
>
> The second group, he says, is made up of 'authorities', 'dominions', and 'powers.' And the third, at the end of the heavenly hierarchies, is the group of 'angels', 'archangels', and 'principalities'.[20]

The three groups each have levels of power (primary, middle and lower order) accorded to them. Ultimately, in the author's view, 'a hierarchy is a sacred order, a state of understanding and an activity approximating as closely as possible to the divine.'[21] On this basis, the Pseudo-Dionysian formula of triadic hierarchies is reflected (albeit corrupted by fallen human nature) by the earthly condition of courts and courtiers, princes and principalities, and their arrangements of tapering authorities structured beneath supreme human leaders. Such structures are given to be divinely instituted realities for which the emanation of light from a single primary source provides the key metaphor:

> And it [i.e. the hierarchical order] is uplifted to the imitation of God in proportion to the enlightenments divinely given to it. The beauty of God – so simple, so good, so much the source of perfection – is completely uncontaminated by dissimilarity. It reaches out to grant every being, according to merit, a share of light and then through a divine sacrament, in harmony and peace, it bestows on each of those being perfected its own form.[22]

Whilst the ideas of *The Celestial Hierarchy* (and chief forebear, the Neo-Platonist theory of *emanation ex deo* conceived by Plotinus, d. 270) had

---

[20] Pseudo-Dionysius, *The Celestial Hierarchy*, trans. Luibheid, 200D–201A, p. 160.
[21] Pseudo-Dionysius, *The Celestial Hierarchy*, trans. Luibheid, 164D, p. 153.
[22] Pseudo-Dionysius, *The Celestial Hierarchy*, trans. Luibheid, 164D, pp. 153–154.

varying influence on medieval thought on light and hierarchy in subsequent centuries, other more practicable engagements with *Timaean* extramission emerged later in the period, which are more indicative of a scholastic approach. Proponents of the Platonist model of visual perception such as William of Conches (*c*.1085–1154) sought to re-establish an extramissive, corporeal understanding of light and vision utilizing natural biological explanation. In his *Glossae super Timaeum Platonis*, William took the corporeal nature of Platonic light to its logical conclusion and advocated that if sight occurred via an interior ray, exterior light and opaque object, a 'medical' description of the biological origins of the interior ray was required. This, he proposed, involved food being digested twice (secondarily in the liver), from which process a 'vapour' was produced, which was transformed by various organ related means into a 'natural virtue', which reaching the heart would become a 'spiritual virtue' and enter the brain. This 'virtue' would be further refined to become an 'airy substance' transmitted from the optic nerve to the eye and emitted from the pupil whenever the soul wished to see.[23]

Aristotelian visuality and colour the 'proper object of sight'

So far the medieval relation of light to colour appears rather one-sided, colour not having been mentioned. It is perhaps significant for the history of the conceptual relationship of light and colour grounded on the *Timaeus* that the text itself was only transmitted to the West via Chalcidius' unfinished translation of *c*.321. This source text finished at paragraph 53b – well ahead of the later fascinating discussion of colours as 'flames given off by bodies' given in *Timaeus* 67c–68d.[24] It is perhaps

---

[23] See William of Conches, *Glosae super Platonem: texte critique avec notes et tables*, ed. Edouard Jeauneau, Textes Philosophiques du Moyen Âge XIII (Paris: J. Vrin, 1965), pp. 236–237.

[24] Plato's *Timaeus* was first translated into Latin during the first century BCE by Cicero, but who only translated paragraphs 27d–47b (ending with the presentation of the significance of vision), arguably to include it in a work of his own on cosmology and to show his mastery of a particularly difficult Greek text. Galen (*c*.129–200 CE) also paraphrased and summarized parts of the *Timaeus*, with a specific interest in the later parts (82a–86a) dealing with the origins of diseases. Galen's paraphrases were the key source for the reception of the Timaeus in the Arab world, where it was translated and commented upon by great scholars such as Avicenna (980–1037) and Averroes (1126–1198). Only much later, through the translation of Marsilio Ficino (1433–1499), did the western world have access to the complete text. See P. E. Dutton, 'Medieval Approaches to Calcidius', in *Plato's Timaeus as Cultural Icon*, ed. G. Reydams-Schils (Notre Dame, IND: Notre Dame

arguable that medieval discourse relating to the Platonist formulation significantly underplays the nature and role of colour in visual perception and the world simply due to its complete absence from writers' knowledge of the *Timaeus*. Moreover, the Aristotelian model of visual perception as an 'intromissive' process outlined in *De anima*, one in which colour is the 'proper object of sight' (*De anima* II. 418b) took longer to be absorbed by the gatekeepers of learning, largely due to its origination from a more suspicious, potentially corrupting pagan writer. *De anima* constituted only a tiny part of the corpus of Aristotelian works transmitted piecemeal to the West, and did not arrive in accessible Latin form until James of Venice's translation of *c*.1150. However, the emergence of a new conceptual relationship between light and colour must be seen to derive from the impact of this text. Most obviously paving the way towards a new understanding, Aristotle rejected the Platonic view of light as a fiery ray emitted by the eyes. In the *De sensu et sensato* section of the *Parva naturalia* he argued:

> It is, to state the matter generally, an irrational notion that the eye should see in virtue of something issuing from it; that the visual ray should extend itself all the way to the stars, or else go out merely to a certain point, and there coalesce, as some say, with rays which proceed from the object. It would be better to suppose this coalescence to take place in the fundament of the eye itself. But even this would be mere trifling. For what is meant by the 'coalescence' of light with light? [...] And how could the light inside coalesce with the light outside it? (438b) For the environing membrane comes between them [i.e. the cornea].[25]

Thus for Aristotle colour, not light, was the primary material of visual perception; visuality a kind of chemical process in which the nature of an object was transmitted via light. Light was therefore, by contrast, an incorporeal medium or receptacle through which images or *species* of objects were delivered into direct contact with the eye – thereby transforming the potential of the eye to perceive into the actuality of perception. Aristotle's intromissive position is made explicit in *De anima*:

UP, 2003), pp. 183–205; a useful summary of the transmission and reception of Plato's Timaeus is Barbara Sattler, *Plato's Timaeus: Translations and Commentaries in the West* <http://www.library.uiuc.edu/rbx/exhibitions/Plato/Pages/Translations.html> [accessed 18 January 2013].

[25] Aristotle, *The Parva naturalia; De sensu et sensibili*, trans. J. I. Beare in *The Works of Aristotle Volume III*, ed. W. D. Ross (Oxford: Clarendon Press, 1908), 438a–b.

the sensitive faculty is such as the sensible object is in actuality. While it is being acted upon, it is not yet similar, but, when once it has been acted upon, it is assimilated and has the same character as the sensible object.[26]

Colour was therefore the 'proper object of sight':

> The object then of sight is the visible: what is visible is colour [...] colour is universally capable of exciting change in the actually transparent, that is, in light; this being in fact the true nature of colour.[27]

The 'chemical' nature of colour that supports the Aristotelian model of perception is outlined in *De sensu et sensato* and derives from his system of elemental transformations via generation and corruption into one another. The model, as it is described in *De generatione*, follows the basic formula of his *Physics*, whereby differences between the different 'species' of colours are to be understood as the product of elemental changes in relation to their contrariety:

> Everything that comes to be or passes away comes from, or passes into, its contrary or an intermediate state. But the intermediates are derived from the contraries – colours for instance, from black and white. Everything, therefore that comes to be by a natural process is either a contrary or a product of contraries.[28]

The wider implications of this theory of elemental interaction included the logical determination that primary material properties can only affect similar properties: 'whiteness could not be affected by line [...] for it is a law of nature that body is affected by body, flavour by flavour, colour by colour' (323b), and that only through contrariety are materials 'susceptible' to the transformations of generation and corruption. This entails ready adaptability in shape by being 'divisible' (328b). From this position of contrariety as the grounding of generation (as caused by susceptibility to contact) Aristotle reached an interesting conclusion for perception:

> For such things can be combined without its being necessary either that they should have been destroyed or that they should survive absolutely unaltered: and their combination need not be a composition, nor merely relative to perception (328b).

---

[26] Aristotle, *De anima* II. 418a.
[27] Aristotle, *De anima* II. 418b.
[28] Aristotle, *Physics* I. 188b.

This conceptually difficult difference between 'combination' and 'composition' implied that a material property (including colour) had an underlying 'fixed' status in relation to its (subjective) perceptions: i.e., that it had its fundamental, independent basis as a property of the elemental nature of a substance, which could then be accessed by 'true' perception of it.[29]

The significance of the four elements and their prime contraries in relation to colour within the natural world emerged further in the *Meteorologica*. In this work Aristotle recapitulated then applied his theory of the elements and their transformations and contrary qualities to particular examples of natural phenomena, including the sea, snow, rainfall, thunder, lighting, whirlwinds, comets, rivers, rainbows, earthquakes, and typhoons. His method was to describe the events and their qualities as products of elemental transformations and the generation or corruption of bodies. Thus patterns of activity are deemed to emerge in terms of various binary oppositions at work between hot and cold, and wet and dry bodies. For instance, in 365b–366a earthquakes are accounted for as 'wind trapped in the earth'; the particular form of 'air' in this case being a 'dry exhalation' as opposed to 'vapour'—the opposing wet variety. Again, in his treatment of rainbows (371b–375b), Aristotle applied an elementally rooted explanation which produced an understanding in terms of three colours:

> These are almost the only colours which painters cannot manufacture: for there are colours which they create by mixing, but no mixing will give red, green, or blue/violet. These are the colours of the rainbow, though between the red and the green a yellow colour is often seen (372a).[30]

---

[29] Such a view of perception, although advocated by Aristotle in the fourth century, anticipated a similar desire for direct conceptual symmetry that can be seen in contemporary scientific reductivism, which insists upon direct correspondences between subjective perceptions and, e.g. neurological events. See, for example, Patricia Smith Churchland, *Neurophilosophy: Toward a Unified Science of the Mind-Brain* (Cambridge, MA; London: MIT Press, 1986).

[30] It is important to remember that the use of Aristotle in translation gives the false impression that ancient Greek colour terminology can be neatly captured by modern English. There has been considerable scholarly interest in the question of how to translate ancient Greek colour terms into other idioms and the cultural, psychological and conceptual problems involved. See especially P. G. Maxwell-Stuart, *Studies in Greek Colour Terminology*, volumes I and II (Amsterdam: E. J. Brill, 1981) for the view that Greek colour terms require a contextual approach, as opposed to that espoused by Irwin who suggests that the relative anomalies in the Greek colour lexicon (which nineteenth-century historians motivated by an

The explanation extends from an inherited position that rainbows are caused by reflection (373a) and that the reflective surface involved is water, specifically: 'water in the process of formation [...] for each of the particles which when condensed forms a raindrop will necessarily be a better mirror than mist' (373b). Moreover, Aristotle states that each of the droplets reflects a single colour, a colour which is magnified by the quantity of drops (we might say akin to an increase in the resolution of a digital picture). Thus the three different colours of the rainbow are accounted for by three apparent facts (374b): (1) Light reflected on a dark surface (like water) or passing through a dark coloured medium produces 'red'; (2) Vision becomes weaker with distance; and, (3) Objects appear darker if seen with weaker vision (caused by distance or infirmity). Therefore, since our sight is weakened when seeing a reflection, there are three stages of this diminishing strength involved in the seeing of rainbows (stages which produce three colours): bright light directly reflected by the dark medium of water produces 'red', while 'green' and 'blue/violet' are produced relative to the further weakening of sight. 'Yellow', Aristotle argued, can also be seen, but is accounted for as a secondary product of the colour contrasts involved and is not due to the primary reflective composition of the rainbow (375a).

The key significance of Aristotle's account of the rainbow for the wider field of medieval colour conceptualization is that it both located colour 'in' the world as a physical property of a body (water) revealed by light, and maintains that, if transmitted by reflection, colour is perceived differently according to the material of the medium by which it is reflected, and the strength of the perceiving eye. Colour, therefore, for Aristotle and his followers, was fundamentally grounded and determined by matter, but what was 'seen' could be co-determined by a combination of the admixtures of elements and their properties in bodies (which are revealed via contrary properties experienced by and through matter: hot, cold, wet and dry), *and* the affects of reflection and subjective visual ability.

> evolution theory explained as national colour blindness) show an overall absence of pure chroma concepts in Greek language and culture. This seems plausible but neglects the evidence of Greek philosophy. There are clearly 'pure' colour concepts defined in the works of Plato and Aristotle – they are simply radically different from our own. See also Eleanor Irwin, *Color Terms in Greek Poetry* (Toronto: Hakkert, 1974).

In *De sensu et sensato* Aristotle also argued that our sensory organs number five, but necessarily correspond to the four elements of material bodies because oral 'taste' is another kind of touching (436b–437a). Hence each type of sense organ corresponds to a different element:

> we must conceive that the part of the eye immediately concerned in vision consists of water, that the part immediately concerned in the perception of sound consists of air, and that the sense of smell consists of fire [*because odour is a 'smoke-like vapour' which arises from fire*] [...] The organ of touch properly consists of earth, and the faculty of taste is a particular form of touch.[31]

Hereafter, the means by which objects are sensed via the watery medium of the eye is the abstract notion of the 'translucent': this being a 'common nature' or 'power' which subsists in bodies in different degrees. Bodies in which the transparent subsist have determinate 'boundaries', and hence surfaces, which entail that 'colour' is at the limit or boundary of a body, and at the limit or boundary of a bodies' translucent power. Outside a determinate body, however, 'translucence' occupies an indeterminate boundary (presumably ultimately occupying the entire sub-lunar region) and so is only seen when a fiery element is present within it, which is in fact what constitutes visible 'light' (439a). 'Colour' (as an object for sensation) is therefore defined as 'the limit of the translucent power in a determinately bounded body' (439b). Thus Aristotle argued that that which produces visible light in air (i.e. the aforementioned fiery element) may also be present in the translucent nature of determinate bodies (or is not when it is 'dark'). This enabled him to posit a key parallel between 'light' and 'dark' (in the indeterminate body of air) and 'white' and 'black' (in determinate bodies):

> Accordingly, as in the case of air the one condition is light, the other darkness, in the same way the colours white and black are generated in determinate bodies (440b).

Colour was, therefore, considered to be the external edge of the material nature of a physical body. Different colours were understood as differing mixtures of material elements ranging between the two extremes of white and black. Each was produced from the 'mixing' together of the white and black elemental parts within bodies. The 'mixing' of the bodies— defined as a 'complete interpenetration' of elemental matter—yielded for

---

[31] *De sensu et sensato*, trans. J. I. Beare, 438b–439a.

Aristotle seven basic natural colours 'in' the world (black, white, yellow or grey, crimson, violet, green and deep blue):

> For there are seven species of each (colours and savours), if, as is reasonable, we regard dun [or grey] as a variety of black (for the alternative is that yellow should be classed with white, as rich with sweet); while crimson, violet, leek-green, and deep blue, come between white and black, and from these all others are derived by mixture.[32]

In effect, Aristotle established a materially-based scale of colour, one that placed green nearer white than blue, and violet closer to black than crimson; i.e., presenting a scale of chromatic development from white to black via green, blue, crimson and violet. The scale was to influence and inspire medieval thinkers and authors in numerous ways. Robert Grosseteste (1175–1253), for instance, in his minor work, *De colores* (the full text of which later appeared in Bartolomaus Anglicus' encyclopaedia, *De proprietatibus rerum*, book 19, compiled *c*.1245), redefined the Aristotelian model to include a concept of brightness, so that the basic colours doubled in number, ascending towards white in one direction and descending towards black in the other:

> Erunt ergo in universo colores sedicim: duo scilicet extremi et hinc inde septem extremis annexi hinc per intensionem ascendentes illunc per remissionem descendentes as in medio in idem concurrentes.[33]

> [There will be, therefore, sixteen colours in the world: two of course of the limit (*i.e. white and black*) and from here thence seven attached to the limits, from here by extension climbing on that side, through abating, descending and concurring into the same middle.]

Grosseteste argued that, 'colour is light embodied with the transparent. In fact, the transparent has two differences: it is indeed transparent—or pure—separate from land, or an impure mixture with earth.'[34] Thus he attempted to synthesize the corporeal Platonic and incorporeal Aristotelian models and traditions regarding light (an approach continued by Roger Bacon) so that the elemental mixtures of colour, regarded as

---

[32] *De sensu et sensato*, trans. J. I. Beare, 442a. Note again that the colour terms used in the Greek do not translate well into English.

[33] Latin quotations from Robert Grosseteste, *De colore*, in *Die Philosophischen Werke des Robert Grosseteste, Bischofs von Lincoln*, Beiträge zur Geschichte der Philosophie des Mittelalters, 9, ed. L. Baur (Münster: Aschendorff Verlag, 1912), pp. 78–79.

[34] 'Color est lux incorpore perspicuo. Perspicui vero duae sunt differentiae: est enim perspicuum aut purum separatum a terrestreitate, aut impurum terrestreitatis admixtione.'

embodied light, could operate within a scale of brightness. This recalibration of the Aristotelian colour theory also helped to insert a chromatic aspect into the legacy of hierarchical and corporeal Christian 'light' theology.

One species of evidence for the 'new' Aristotelian conceptual relationship forged along these lines, and found in medieval culture at large, is provided by fourteenth-century books of arms. For instance, Johannes de Bado Aureo's *Tractatis de Armis* (written sometime soon after 1394; a date established by the author's dedication of the work to the recently deceased Queen Anne) sets out to explain the symbolism and significance of armorial devices and the practice of armorial bearings from antiquity to his own day.[35] Whilst Bado's writing is, as one would expect, informed by a tradition of books of chivalry, chronicles and heraldic writings stemming principally from Ramon Llull's seminal *Libre Del Ordre De Cauayleria* (c.1280), as well as the armorial aspect of romance writing from the twelfth century onwards, Aristotelian colour theory has also, and influentially, found its way into the mix. The first section of Bado's text is devoted entirely to the significance of armorial colours, regarded independently of the devices or forms they are colouring. Bado's view was that the primary mode for significance of arms was colour. Having dedicated his work and referred it to his authority ('mei magistri Francisci de Foveis ies') Bado states his initial intention: 'Primo differentias colorum ponam, et quis eorum dignior vel nobilior inveniatur' ('First, I will put forward the different colours, so that the more worthy or more noble of them may be discovered').[36] He then

---

[35] This is one of a family of heraldic texts going back to the Anglo-Norman *De Heraudie* (composed c.1300). The first that evidences Aristotle's colour theory in the treatment of armorial colours is the highly influential *De Insigniis et Armis* of Bartolo di Sasso Ferrato (c.1355) to which Johannes de Bado refers (see below). For details of other medieval heraldic texts see Rodney Dennys, *The Heraldic Imagination* (London: Barrie and Jenkins, 1975) pp. 59–71; for texts of Bado's *Tractatus* and Bartolo's *De Insigniis* see Evan Jones, ed., *Medieval Heraldry: Some Fourteenth Century Heraldic Works* (Cardiff: William Lewis, 1943), pp. 3–21 & pp. 95–212.

[36] The opening of the text reads: 'Quoniam de armis multociens in clepeis depictis singula discernere et describere inveniatur difficile, ad instantiam quarundam personarum et specialiter ad instantiam Dominae Annae quondam Reginae Angliae hunc libellum compilavi, sequens in parte dogmata ac traditiones excellentissimi Doctoris et Praeceptoris mei magistri Francisci de Foveis, omnipotentis Dei nomine primitus invocato sub hac forma procedam. Primo differentias colorum ponam, et quis eorum dignior vel nobilior inveniatur.' Johannis de Bado, *Tractatus de Armis*,

arranges hierarchically the colours employed in heraldry following an Aristotelian scale, in order to answer an important ideological question for chivalry: 'Which colour is the most honourable?'[37] In answering this dilemma, Bado assesses the colours of fourteenth-century heraldry according to their status as primary, mediary or sub-mediary; their relative honour determined by the amount of elemental mixing gone into their creation. Thus Bado's opines that white is the highest colour in terms of honour, closely followed by black – as opposed to Bartolo's contention that black was the lowest colour.[38] (Red and blue are superior to green (as sub-mediary) in all fourteenth-century accounts of armorial colours I have seen):

> Colores principales secundum se sunt color albus et niger; colores vero medii sunt azoreus, aureus, et rubeus; colores autem submedii sunt color viridis et alii similes si inveniantur. Et ratio mea est quare dico illum colorem submedium, quia non potest aliquo modo fieri ex duobus coloribus principalibus, scilicet, albo et nigro sed sit dumtaxat ex duobus coloribus mediis, scilicet ex colore azoreo et colore aureo adinvicem mixtis.[39]

> [After the principal colours which are white and black; the truly mediary colours are blue, gold and red; on the other hand, the sub-mediary colours are green and any others if obtained similarly. And the reason why I say this is that the sub-mediary colours are not composed from the two principal colours, which are of course white and black, but as far as it applies from admixtures of two mediary colours, which are of course the colour blue and the colour gold.]

---

ed. Jones, p. 95. Jones's edition has two variants of De Bado's tract, one based on similar copies: London, British Library, Add. MS 37526 and Add. MS 29901, and a slightly different version from Add. MS 28791. All three belong to the fifteenth century.

[37] The question is put explicitly in the Welsh version of the Tractatus, *Llyfr Dysgread Arfau*, composed by Bishop Trevor (*c*.1405–10): 'Pa liw ysydd anrrydeddusaf?' (Jones, p. 8).

[38] 'Colores autem medii sunt nobiliores vel minus nobiles secundum quod plus vel minus appropinquant albedini vel nigredini. Istud videtur dictum Aristotelis in libro suo, De Senso et Sensato [...] color albus est nobilior quia magis appropinquat luci; color niger est infimus quia magis appropinquat tenebris.' ('the colour white is nobler because it greatly approaches light; the colour black is least because it greatly approaches darkness.') Bartolo di Sassoferrato, *De Insignis et Armis*, ed. Jones, p. 247, para. 27.

[39] Johannis de Bado, *Tractatus de Armis*, ed. Jones, pp. 99–100.

A distinct aspect of Aristotelian influence on medieval thought about visuality, therefore, was that white or 'whiteness' could occupy the conceptual space previously held by light, by demonstrating the value of purity via an absence of material 'contamination'. The Aristotelian chromatic spectrum could be employed therefore as the basis for a different mode of hierarchical thinking about visuality, one corresponding analogously to the hierarchal system governed by the traditional theology of extramissive light. By the time of Johannis de Bado's *Tractatus*, the synthesis of belief with visuality had developed a precise means of redefining courtly society through a chromatic hierarchical identification system, one based upon a part secular, part spiritual treatment of perception that operated within particular contexts, e.g., armory. Interestingly enough, taking an Aristotelian approach to armory also entailed the view of green (*vert*) as *least* honourable of the colours used in heraldry. Thus Bado wrote of green: 'Quidam tamen addunt alium colorem, scilicet viridem colorem, qui color, ut ego credo, initium habuit ab aliquo milite histrione vel gaudente,' ('Some would add yet another colour to those above noted, namely, green, which as some suppose was borne first of all by a minstrel, and afterwards he was granted arms').[40] According to statistical data such as that assembled by G. R. Samson in 2000,[41] there was, tantalizingly, a relative paucity of green used in armorial displays during the early period of heraldic display. Samson's data for the fourteenth century yields a frequency of usage pattern placing *gules* (red) as the most frequently used colour, followed by *argent* (silver/white), *or* (gold/yellow) then *azure* (blue) before *sable* (black). The least used colour by a considerable margin was *vert* (green). The attitude, perhaps based in part on the application of the Aristotelian colour scale to chivalry, changed during the fifteenth century. Nicholas Upton's *De studio militari* draws especial attention to the point. Writing in his introduction, Upton pointedly distances himself from the indiscretions of his youthful beliefs relating to green:

> Olim, in annis meis juvenilibus, scripti in hac materiale nimium sompniando: in qua quidem scriptura sateor me multipliciter erase, ut in

---

[40] *Tractatus de Armis*, ed. Jones, pp. 99–100.
[41] G. R. Samson, 'Historical Trends in the Deployment of Tinctures', *The Coat of Arms*, n.s., 13 (2000), 271–77.

dampnando colorem viridem, ac multa alia posuisse que sunt veritati contraria: que jam ex certa mea scientia revoco.[42]

[Formerly, in my youth I wrote on this matter in too dreamy a manner, and in my writing I must confess to have made many errors, as in condemning the colour green, and I have stated many other matters which are contrary to the truth: which now I revoke from my knowledge.]

Upton went on to write that he would like to burn his early mistakes and now proposes to correct them. To explain his vehement change of heart, Dennys points out that 'disparaging remarks about the colour green in arms would not have been well received by his master, the Earl of Salisbury, the second quartering in whose arms was that for Monthermer, *Gold an Eagle displayed vert*'.[43]

Suffice to say of this vast topic, in broad terms the inferiority of *vert* and hierarchy of armorial colours in fourteenth century books of arms may also sound an interestingly note for the interpretation of romance texts from the same period, in particular *Sir Gawain and the Green Knight*.[44]

Conclusion

This brief overview of the intellectual and phenomenological terrain available to medieval thought and its imagination relating to light and colour suggests that the relationship changed subject to increasing

---

[42] Upton's *Studio de militari* is believed to have been written before 1446, since it is dedicated to Humphrey, Duke of Gloucester, who died on 28 February of that year. For a partial edition and biographical material, see Francis Pierrepont Barnard ed, *The Essential Portions of Nicholas Upton's De Studio Militari: Before 1446, trans John Blount c.1500* (Oxford: Clarendon Press, 1931); also Dennys's lengthy treatment in *The Heraldic Imagination*, pp. 76–82, which lists numerous MS copies of the text. The quotation used here is from the Bernard edition, p. ix.

[43] Dennys, p. 78.

[44] I have discussed the question of the 'greenness' of the Green Knight in relation to fourteenth-century armorial theories at length in my recently submitted PhD thesis, 'Colour, Seeing, and Seeing Colour in Medieval Literature'. In summary this argues that the Green Knight, if read armorially, should be done so in relation to fourteenth-century book of arms under the influence of Aristotelian colour theory, as opposed to differently orientated early Renaissance works such as Jean Courtois's influential *Le blason des coleurs en armes* (c.1414), a work referred to for instance by Derek Brewer in his useful overview of the colour green in relation to Sir Gawain and the Green Knight, cf. *A Companion to the Gawain-Poet*, ed. Derek Brewer and Jonathan Gibson, Arthurian Studies, 38 (Cambridge: D.S. Brewer, 1997), pp. 181–90 (pp. 186–7).

assimilation of Aristotelian philosophy from the twelfth century onwards. Such a situation had certain logical consequences: colour became more available as a value in and of itself, and able to be hierarchically ordered for the purposes of abstract perceptual judgement. Fourteenth-century armorial texts evidence this effect at its most explicit – ordering hues in relation to a spiritual ideology in which 'material' purity was of paramount importance. White, it would seem, was the new light.

THE UTILITY OF THE *LUX-LUMEN* DISTINCTION IN ROGER BACON'S THOUGHT

David M. Barbee

With a purported emphasis on empiricism, Roger Bacon is considered by some to be an early precursor of modern science.[1] While this claim may be true, it is a radically reductionist view of the thirteenth century

---

[1] Bacon seems to view experiments as a valuable part of the scientific process. He candidly confesses, 'without experience nothing can be sufficiently known'. Further, he places experience in contrast with reasoning. 'Reasoning draws a conclusion and makes us grant the conclusion, but does not make the conclusion certain, nor does it remove doubt so that the mind discovers it by the path of experience; since many have arguments relating to what can be known, but because they lack experience they neglect the arguments, and neither avoid what is harmful nor follow what is good', Bacon writes. [See Roger Bacon, *Opus maius*, trans. Robert Belle Burke, 2 vols (Philadelphia: University of Pennsylvania Press, 1928), 6.1]. This has led Brian Clegg and others to describe Bacon as the first scientist. Clegg's book is a popular account of Bacon's life, but more academic authors such as William Whewell and Andrew Dickson White have argued for Bacon as a precursor of modern science. However, Bacon still relies on authorities on many points in the *Opus maius*. It is best to judge Bacon within his own historical context, as David Lindberg comments, 'it is obvious that Bacon's aim as a scholar was not to anticipate the future, but to respond to the past; it could not have entered his mind that his mission in life was to anticipate some future methodology, and it follows that his failure to do so cannot be taken as a defeat.' Despite this fact, modern researchers still largely study Bacon for his views on various sciences. See Brian Clegg, *The First Scientist: A Life of Roger Bacon* (New York: Carroll & Graf, 2003), David C. Lindberg, *Roger Bacon and the Origins of Perspectiva in the Middle Ages: A Critical Edition and English Translation of Bacon's 'Perspectiva' with Introduction and Notes* (Oxford: Clarendon Press, 1996), pp. lii–lvii, and Jeremiah Hackett, ed., *Roger Bacon and the Sciences* (Leiden: E.J. Brill, 1997). For a summative assessment of views of Bacon as a scientist, see Jeremiah Hackett, 'Experience and Demonstration in Roger Bacon: A Critical Review of some Modern Interpretations', in Alexander Fidora and Matthias Lutz-Bachmann, eds, *Erfahrung und Beweis: Die Wissenschaften von der Natur im 13. und 14. Jahrhundert / Experience and Demonstration: The Sciences of Nature in the 13th and 14th Centuries* (Berlin: Akademie Verlag), pp. 41-58. On Bacon's reception generally, see Amanda Power, 'A Mirror for Every Age: The Reputation of Roger Bacon', *English Historical Review*, 121.492 (2006), 657-92.

Franciscan thinker. In fact, Bacon is very nearly a polymath. In his *Opus maius*, Bacon sets out to formulate a universal theory of knowledge based upon theology. In doing so, he incorporates parts of the traditional liberal arts taught during the Medieval period as well as optics, mathematics, and moral philosophy. Bacon very clearly had varied interests that extended well beyond the confines of modern science.

One field in which Bacon's divergent interests converge is in his conceptualization of the metaphysics of light. It is this notion that effectively serves as a unifying theory behind the Baconian quest for a universal science. As a philosopher of nature, Bacon developed a theory of optics; as one who fancied himself a theologian, he believed that the knowledge attained through his study of optics had direct relevance for Christian theology.[2] In beginning to construct his metaphysics of light, Bacon exploits a metaphor of light to aid in explaining his illumination theory of human knowledge. Next, he believes that God creates the universe in a fashion similar to the way light extends out from light. Finally, Bacon utilizes light as a model for motion that allows him to prove the existence of God as first cause. The use of 'light' is found on three levels in Bacon's thought. At the center of his metaphysics of light is a distinction between *lux* and *lumen* that serves as an explanatory model for his epistemology and his cosmology, as well as other human sciences.

First, Bacon's methodology must be illuminated. Bacon is in harmony with the Augustinian notion that theology is the queen of the sciences and all other sciences are mere handmaidens. He maintains an essential unity between the knowledge revealed in theology and that which is gained through philosophy and the sciences. 'Therefore philosophy is merely the unfolding of the divine wisdom by learning and art,' Bacon asserts. He continues, 'Hence there is one perfect wisdom which is

---

[2] Although Bacon exerted considerable energy in the attempt to reform theological studies, he was never technically a theologian. Bacon received his M.A. around 1240 from Oxford or Paris. After receiving this degree, he began to lecture on the works of Aristotle as a member of the Faculty of Arts in Paris. He never graduated to the Faculty of Theology. Bacon gave up his post and moved back to Oxford around 1251. At this point, he seems to have involved himself in reading and experimentation, but there is no definitive proof for his actions during this period until he joins the Franciscan Order around 1257. Bacon wrote no biblical commentaries or other theological works, or at least none that we know about, but he composed books that sought to reform theological education in the universities across Christendom by including languages and natural sciences. On Bacon's attempts at reform, see Amanda Power, *Roger Bacon and the Defence of Christendom* (Cambridge: Cambridge University Press, 2012).

contained in the Scriptures, and was given to the saints by God; to be unfolded, however, by philosophy as well as by canon law.'[3] Bacon cites St Paul, St Augustine, Aristotle, Plato, and Cicero to the effect that the wisdom of philosophy ultimately reduces to divine wisdom because God is the ultimate source of both bodies of knowledge.[4] This line of argument is extended when Bacon concludes that whatever truth a person may find, it can also be found in the Bible.[5] Put more negatively, Bacon contends that 'every consideration of a man that does not belong to his salvation' meaning anything that is not contained in the Bible, which is tantamount to heresy for Bacon, 'is full of blindness, and leads down to the darkness of hell'.[6]

In viewing knowledge in this fashion, Bacon is attempting to navigate between two extremes—an independent science inspired by Aristotle that seemingly contradicts the Bible and a fundamental fideism that is dismissive of all other sciences except theology.[7] He maintains that all knowledge is useful in helping a Christian interpret the Bible and, as such, ought to be pursued. Even the science of optics, Bacon will claim, is important because:

> in divine scripture, nothing is dealt with as frequently as matters pertaining to the eye and vision [...] and therefore nothing is more essential to [a grasp of] the literal and spiritual sense than the certitude supplied by this science

---

[3] Bacon, *Opus maius*, 2.14.

[4] Ibid., 2.6.

[5] Ibid., 2.1. Bacon's claim on this point is revelatory of his attitude toward the changes he observed in the study of theology. Bacon was a critic of the increasing use of philosophy in the exposition of theology insofar as he believed that philosophy was beginning to supplant biblical exegesis as the appropriate theological method. Beryl Smalley aptly depicts Bacon's attitude toward contemporary developments in the theological science. Smalley describes Bacon as 'a rebellious reactionary, or a reactionary rebel' who clung to the Alexandrian approach to the Bible and the exegetical methodology of Hugh of St Victor. See Roger Bacon, *Opus minus* in *Fr. Rogeri Bacon opera quædam hactenus inedita*, ed. J.S. Brewer, Rolls Series (London: Longman, Green, Longman and Roberts, 1859; repr. Nendelin: Kraus, 1965), pp. 322–23 and Beryl Smalley, *The Study of the Bible in the Middle Ages* (Notre Dame, IN: Notre Dame University Press, 1978), pp. 329–30.

[6] Ibid.

[7] David C. Lindberg, 'The Medieval Church Encounters the Classical Tradition: Saint Augustine, Roger Bacon, and the Handmaiden Metaphor', in *When Science and Christianity Meet*, ed. David C. Lindberg and Ronald Numbers (Chicago: University of Chicago Press, 2003), p. 22.

[...] For example, when it is said 'Preserve me, oh Lord, as the pupil of your eye' it is impossible to know God's meaning in this phrase unless we first consider how the preservation of the pupil is achieved [...] For when something is set forth as an example and similitude, that which is exemplified cannot be understood unless the nature of the example is grasped.[8]

Stewart Easton rightly comments, 'It was one of the primary tasks of the scientific investigator to complete, and, as Bacon emphasized, to confirm the knowledge given in summary or occult form in the Scriptures, and in the writings of the Saints and Fathers of the Church, and of anyone who had had direct access to the revealed knowledge.'[9] It is for this sort of utilitarian reason that Bacon sets out to construct his universal science.

Underlying Bacon's methodology is the epistemological assumption that it is necessary for God to reveal knowledge to human beings. Bacon contends, 'Although in some measure the truth may be said to belong to the philosophers, yet for possessing it the divine light first flowed into their minds, and illumined them from above.'[10] Amongst a number of authorities, he cites a portion of the second book of Augustine's *De Doctrina Christiana*. In this portion, the bishop of Hippo argues that all knowledge of metaphysics and the science of creation are in the Bible

---

[8] Roger Bacon, *Perspectiva*, III, d. 3, c. 1 in Lindberg, *Roger Bacon and the Origins of Perspectiva in the Middle Ages*. On the connection between the metaphor of seeing and religion, see Dallas G. Dennery II, *Seeing and Being Seen in the Later Medieval World: Optics, Theology and Religious Life* (Cambridge: Cambridge University Press, 2009).

[9] Stewart Easton, *Roger Bacon and His Search for a Universal Science: A Reconsideration of the Life and Work of Roger Bacon in the Light of His Own Stated Purposes* (Oxford: Basil Blackwell, 1952), p. 172. From the *Opus maius*, it is unclear whether Bacon would have endorsed the study of sciences in their own right for the improvement of humankind. Further, Bacon does not claim that the *only* reason for studying the sciences is for the development of theology. One must keep in mind his purpose in composing the *Opus maius*. Bacon is trying to persuade his readers to include the sciences in the theology curriculum. Naturally, then, Bacon has to show their utility for theology. Henri de Lubac observes two complimentary trends amongst medieval theologians, 'One of these tended to make each of the profane or "liberal" disciplines serve to elaborate the "spiritual disciplines", which is to say, more than anything else, the explication of Scripture. The other idea tended to go further, finding in Scripture the source and summit of all knowledge, profane as well as sacred.' There does not appear to be a reason for believing that Bacon ignored the good that science can bring society, but this was not necessarily his primary concern. See Henri de Lubac, *Medieval Exegesis*, I: *The Four Senses of Scripture*, trans. Mark Sebanc (Edinburgh: T&T Clark, 1998), p. 41.

[10] Bacon, *Opus maius*, 2. 5.

with the implication that all knowledge ultimately flows from God and can only be found perfectly in God.[11] More forcefully, Bacon claims 'the wisdom of philosophy was totally revealed by God and given to philosophers, and God himself [...] illuminates the human soul in all wisdom'.[12] He explains that this illumination was necessary because humans simply could not acquire the principles of science and arts without assistance. This also shows that there is, in fact, a universal science.[13] Internal illumination was granted to the patriarchs and prophets because even if they had knowledge of the arts and sciences, experimental science 'does not give full attestation in regard to things corporeal owing to its difficulty, and does not touch at all on things spiritual'.[14] For Bacon, revelation is absolutely necessary for knowledge of whatever kind, whether it be religion or the seminal techniques of the arts and sciences.

The more mature Bacon explains his theory of illumination by resorting to the Aristotelian distinction between the active intellect and the potential intellect.[15] He elaborates on this distinction in a lengthy passage:

[11] Ibid., 2. 1. He also cites Ambrose on Colossians, 'All knowledge of the science above and of the creation beneath is in him who is the head and author, so that he who knows him should seek nothing beyond, because he is the perfect virtue and wisdom' with the same result, but Bacon is mistaken regarding the authenticity of this authority. A similar phrase appears in Herveus Burgidolensis' eleventh century commentary on Colossians, but it is not credited to Ambrose. It also appears in Gratian, but this time Ambrose is specifically cited. See Herveus Burgidolensis, *Commentaria in Epistolas Divi Pauli* in *PL*, 181. 1328B and Gratian, *Concordia Discordantium Canonum* in *PL*, 187. 203C–204A. The commentaries on Paul that went under the authorship of Ambrose are now credited to Ambrosiaster. It is unclear, though, precisely what source Bacon relied upon for his citation.

[12] Roger Bacon, *Opus tertium* in *Fr. Rogeri Bacon opera quædam hactenus inedita*, p. 74 (see n. 5 above); compare with Bacon, *Opus maius*, 2. 6.

[13] Bacon, *Opus maius*, 2. 9.

[14] Ibid., 6. 1.

[15] The relationship between the active intellect and the passive intellect is one that changes throughout the course of Bacon's career. Initially Bacon taught that the active intellect was part of the human soul. It was only later in life, in the midst of an outcry against Averroism and under the influence of Alfarabi and Avicenna, that Bacon identifies the active intellect with God. As Theodore Crowley demonstrates, this created a problem in Bacon's concept of the soul, as the properties of the active intellect in Bacon's earlier iteration of the soul must be transferred to the passive intellect. Bacon gets around this difficulty by positing a storehouse of actual knowledge contained within the passive intellect that is inaccessible because of

The human soul is called possible by (philosophers), because it has of itself capacity for sciences and virtues and receives these from another source. The active intellect is the one which flows into our minds, illuminating them in regard to knowledge and virtue, because although the possible intellect may be called active from the act of understanding, yet in assuming an active intelligence, as they do, it is so called as influencing and illuminating the possible intellect for the recognition of truth.[16]

Bacon continues on in his argument citing Arabic Aristotelians such as Alfarabi and Avicenna in order to prove that God is the active intellect by virtue of his distinction from the human soul, his incorruptibility, his omniscience, and his actuality. He likens God to light on a few different occasions in this argument. First, he sides with Avicenna who explains the active intellect's relationship to the passive intellect as similar to 'the light of sun to colors' and 'the sun driving away darkness from colors'. In the second instance, Bacon cites Augustine who refers to God as the principal epistemological influence, 'just as to the sun is ascribed the flow of light falling through the window'.[17]

While this appears to be a superficial metaphor of light, Bacon is actually participating in a significant philosophical discussion within the scientific community of the middle ages that alludes to a universal causal relationship. This discussion centers on a distinction made between *lux* and *lumen*. Avicenna distinguishes between three different categories within light: *lux* is the fiery quality, something like fire or the sun possesses that is perceived when a medium intervenes; *lumen*, for Avicenna, is that which radiates out from these luminous bodies and falls upon other bodies, causing them to be seen. Finally, Avicenna also

---

human embodiment, but which is retrieved when a sensible object is encountered. In the *Opus tertium*, Bacon limits the active intellect to God primarily and to the angels in a secondary fashion. Obviously, in the main body of this work, I am concerned with Bacon's later position. See Roger Bacon, *Questiones supra undecimum prime philosophie Aristotelis* in *Opera hactenus inedita Rogeri Baconi*, ed. Robert Steele (Oxford: Clarendon Press, 1909), fasc. 7, 110. 1–18 and 15. 27–16. 15, Roger Bacon, *Questiones supra libros octo physicorum Aristotelis* in *Opera hactenus inedita Rogeri Baconi*, ed. Robert Steele (Oxford: Clarendon Press, 1909), fasc. 13, 11. 35–12. 2, Bacon, *Opus tertium*, Rolls Series, 74, and Theodore Crowley, *Roger Bacon: The Problem of the Soul in his Philosophical Commentaries* (Louvain-Dublin: Éditions de l'Institut supérieur de philosophie, 1950), pp. 163–90.

[16] Bacon, *Opus Maius*, 2. 5.

[17] Ibid. Although Bacon takes some liberty with the text, he appears to be referring to Augustine's comments on Psalm 119:73. See Augustine, *Enarrationes in psalmos*, ed. E. Dekkers and J. Fraipont, CCSL, 40 (Turnholt: Brepols, 1956), 118.18.4.

establishes a category called *radius* or *radiositas* which consists of 'that which appears around bodies [...] as though it were something emanating from them'.[18] Alhacen followed Avicenna in making distinctions between *lux* and *lumen*, although the terminology in the Latin editions of his works obfuscated these distinctions.[19] Later writers, including Averroes and Albertus Magnus, employed this division to speculate on the propagation of light and the nature of the medium that carries light and color to the viewer. David Lindberg comments upon the influence of Alhacen and Avicenna,

> The themes treated by Avicenna and Alhazen underwent continued development in the West. Avicenna's distinction between *lux* and *lumen* was widely employed, though (with the encouragement of Alhazen) it was widely ignored. It was agreed by virtually everybody who touched upon the matter, however, that both light (usually *lux*) and color propagate their forms or likenesses through transparent media to observers.[20]

Drawing upon the work of his predecessors, Bacon clarifies the difference between *lux* and *lumen* in his thought by quoting Avicenna's *De Anima* when he comments,

> *Lux* is a quality of a luminous body, such as a fire or a star; but *lumen* is that which is multiplied and generated from that *lux* and which is produced in air and other rare bodies, which are called media because species are multiplied by their mediation.[21]

---

[18] *Avicenna Latinus: Liber de anima seu sextus de naturalibus*, I–II–III, ed. S. Van Riet (Louvain and Leiden: E.J. Brill and Peeters, 1972), pp. 170–72.

[19] David C. Lindberg, *Studies in the History of Medieval Optics* (London: Variorum, 1983), I.357. For instance, in Book I of the Latin translation of *Opticae thesaurus*, the text states that '*lux*' extends through diaphanous bodies according to straight lines. If the distinction between *lux* and *lumen* is to be maintained, this should have been *lumen*. See Alhacen, *Opticae thesaurus* (Basel, 1572), 1. 17.

[20] Ibid.

[21] Roger Bacon, *On the Multiplication of Species*, 1.1.34–36 in David C. Lindberg, *Roger Bacon's Philosophy of Nature: A Critical Edition, with English Translation, Introduction, and Notes of De multiplicatione specierum and De speculis comburentibus* (Oxford: Clarendon Press, 1983). Bacon treated the subject matter of *De multiplicatione specierum* on several occasions and composed a few strikingly similar texts that have been given comparable titles. I am using this version of *De multiplicatione specierum* rather than that contained in Bridges' edition of the *Opus maius* because Lindberg has painstakingly edited the text and provided a valuable critical apparatus that is lacking in the Bridges edition. For the textual history of *De*

Before Bacon's statement can be properly understood, one has to understand what Bacon means by the word 'species'. He first tries to explain the word with a list of synonyms such as 'the similitude of the agent', 'image', [...] 'idol', 'simulacrum', 'phantasm', 'form', as well as a few other terms. In a more straightforward way, Bacon defines 'species' as 'the first effect of any naturally-acting thing'. In the *lux-lumen* distinction just mentioned *lumen* is the species of *lux* because it is projected out from *lux* as an effect. Further, *lumen* is a species of *lux* in regard to sense and intellect because it is detected by human sense and because it also imitates or is similar to *lux*, although it is a diminished form of the substance of *lux*. He concludes that the 'species produced by an agent is similar to the agent in nature, definition, specific essence, and operation'.[22] Sebastian Vogl aptly summarizes Bacon's concept of species,

> Therefore, Bacon understands by 'species' something resulting from an active cause, namely at first that what is effected by the active cause by virtue of its own nature and has therefore a resemblance to it.[23]

Bacon's notion of a species as a force moving out from an object will prove critical for his optical theory. There were three regnant optical theories during Bacon's time that had been inherited and adapted from antiquity. The Euclidean theory, also referred to as the 'extramission' theory and supported by Ptolemy, maintained that rays extend out from the eye in generally straight lines, unless refracted or reflected, until they come upon an object at which point information is relayed back to the eye.[24] In essence, this was a mathematical attempt to explain depth

multiplicatione specierum, see David C. Lindberg, *Roger Bacon's Philosophy of Nature*, pp. xxvi–xxxv and pp. lxxv–lxxx.

[22] Ibid., 1.92–94.

[23] Sebastian Vogl, 'Roger Bacons Lehre von der sinnlichen Spezies und vom Sehvorgange', in *Roger Bacon*, ed. A.G. Little (Oxford: Clarendon Press, 1914), p. 208.

[24] Euclid, *Optica* in *Opera Omnia*, vol. VII, ed. I. L. Heiberg and H. Menge, (Leipzig: B. G. Teubner, 1895), p. 1 and Ptolemy, *L'Optique de Claude Ptolémée dans la version latine d'après l'arabe de l'émir Eugène de Sicile*, ed. Albert LeJeune, Collection de Travaux de l'Academie Internationale d'Histoire des Sciences, no. 31 (Leiden: E. J. Brill, 1989), II, 22–25. See also Albert LeJeune, *Euclide et Ptolémée: Deux stades de l'optique géométrique grecque* (Louvain: Bibliothèque de l'Université, Bureaux du "Recueil", 1948) and A. Mark Smith, *Ptolemy's Theory of Visual Perception: An English Translation of the Optics with Introduction and Commentary*, (Philadelphia: The American Philosophical Society, 1996 [=*Transactions of the American Philosophical Society* 86, part 2]).

perception. The second theorem, advanced by Lucretius and Epicurus as well as in a slightly different form by Aristotle, held that atoms emanated out from an object in the same shape as the originating object before entering the eye. This was an attempt to explain vision physically.[25] The third hypothesis, defended by Galen, believed that the visual spirit goes out from the eye, changing the surrounding air into an extension of the optic nerve and thereby allowing the medium to carry the object back to the eye. The Galenic theory was primarily concerned with the anatomy of the eye and the physiology of sight.[26] These remained the three main dividing lines in discussion of optics in the medieval period. In his attempt to weld these views together, Bacon strikes a compromise in his optical theory by arguing that species emanate from the eye as well as from a visible object.[27] Species from the seer and the object meet and visual species return to the crystalline humor in the eye where vision begins.

The propagation of species as illustrated in the *lux-lumen* distinction also becomes the model by which Bacon understands creation and motion. In his *Compendium of the Study of Theology*, Bacon perhaps makes a greater distinction between God and creation than he does between *lux* and *lumen*, but there is enough similarity for him to comment, 'there can be a wholly absolute difference among some things and yet [at the same time] there is some agreement in relation, as between the Creator and a creature, for they share in nothing absolutely common, but yet a creature does have a relation to the Creator and is both a sign of

---

[25] Lucretius, *De rerum natura*, in Loeb Classical Library, ed. W. H. D. Rouse and rev. Martin Ferguson Smith, Loeb Classical Library (Cambridge, MA: Harvard University Press, 1975), 4. 54–61, Epicurus, 'Letter to Herodotus', in Diogenes Laertius, *Lives of Eminent Philosophers*, in Loeb Classical Library, ed. R.D. Hicks, 2 vols (Cambridge, MA, 1972), II, 10. 48–9, and Aristotle, *De anima*, II. 7 and *De sensu*, 2–3 both in *The Complete Works of Aristotle*, vol. I, ed. Jonathan Barnes (Princeton: Princeton University Press, 1995).

[26] Galen, *De usu partium*, 2 vols., trans. Margaret Tallmadge May, Cornell Publications in the History of Science, (Ithaca, NY: Cornell University Press, 1968), Book 10. See also Rudolph Siegel, *Galen on Sense Perception* (Basel: Karger, 1970), pp. 40–126.

[27] For a more extensive and nuanced discussion of Bacon's view in regard to these previous positions, see David C. Lindberg, 'Light, Vision, and the Universal Emanation of Force', in *Roger Bacon and the Sciences*, pp. 243–75 and idem, 'The Science of Optics', in *Studies in the History of Medieval Optics*.

Him and [His] effect.'[28] As active intellect, as *lux*, God creates the universe. This is precisely why Bacon can claim that nature is the instrument of divine operations.[29] Bacon alludes to I Corinthians 13. 12 in which St Paul writes 'Now we see but a poor reflection as in a mirror; then we shall see face to face. Now I know in part; then I shall know fully, even as I am fully known.' Interestingly, Bacon interprets this passage to mean that God is now seen in the world through species that have been distorted by numerous reflections.[30] According to Bacon, a person can only comprehend universals through species.[31] It is precisely this belief that allows Bacon to claim that 'the whole aim of philosophy is that the Creator may be known through the knowledge of the creature'.[32] Erich Heck is correct to state that, for Bacon, 'Nature is the reflection of God, in it, God really is seen, even if in broken, weaker radiations.'[33]

---

[28] Roger Bacon, *Compendium of the Study of Theology*, ed. and trans. Thomas Maloney, Studien und Texte zur Geistesgeschichte des Mittelalters (Leiden: E. J. Brill, 1988), 6. 23–25.

[29] Bacon, *Opus tertium*, Rolls Series, 100.

[30] Roger Bacon, *Un fragment inédit de l'opus tertium de Roger Bacon précédé d'une étude sur ce fragment*, ed. Pierre Duhem (Quaracchi: Collegio St Bonaventure, 1909), p. 96.

[31] Roger Bacon, *Communia naturalium* in *Opera hactenus inedita Rogeri Baconi*, fasc. 2, 104. 7–17 (see n. 15 above).

[32] Bacon, *Opus maius*, 2. 7.

[33] Erich Heck, *Roger Bacon: Ein Mittelalterlicher Versuch Einer Historischen und Systematischen Religionswissenschaft* (Bonn: H. Bouvier u Co., 1957), p. 138. This is particularly true given Bacon's conceptualization of semiotics. For Bacon, the importance of a sign is placed on the intention of the being creating the sign, not necessarily the sign itself. As Bacon describes it, the sign is not the thing signifying, but rather the being who makes the sign. In this instance, it is less significant that all of creation is a sign of God which can be built into philosophy, but that God intends for this to be so. Creation ultimately functions in a similar manner to language. Irene Rosier-Catach comments regarding language, 'Language is for man an instrument, a means.' In the same way, creation is a communicative tool for God in Bacon's thought. See Roger Bacon, *Summa de sophismatibus et distinctionibus* in *Opera hactenus inedita Rogeri Baconi*, fasc. 14, 153. 19–154. 21 (see n. 15 above) and Irene Rosier-Catach, 'Roger Bacon and Grammar', in *Roger Bacon and the Sciences*, 67–102. On Baconian semiotics, see Jan Pinborg, 'Roger Bacon on Signs: A Newly Recovered Part of the Opus maius', in Jan P. Beckmann et al., eds, *Sprache und Erkenntnis im Mittelalter: Akten des VI. Internationalen Kongress für Mittelalterliche Philosophie der la Société internationale pour l'étude de la philosophie médiévale, 29 August-3 Septembre 1977* (Berlin: Walter de Gruyter, 1981), pp. 403-

Bacon is straightforward in regard to his use of light as a model for the motion of species. He explains that humans use the words 'radiant' and 'rays' to describe the manner in which species multiply because 'the multiplication of light is more apparent to us than the multiplication of other things, and therefore we transfer the [terminology of the] multiplication of light to the others.'[34] Bacon clarifies Aristotle's opinion and contends that light is only different from the species of other sensibles because it moves in a shorter and less comprehensible time.[35] He believes that authors of works on vision and light 'can and ought to be applied to the other senses—and not only to the senses, but to all the matter of the world altered by species and powers of all agents whatsoever. And therefore the entire action of nature and the generation of natural things take their bases and principles' from these optical theorists and their texts.[36] Eventually, Bacon will use light as a model for the corruption of species as the direction of species can be modified when they pass through things just like light is distorted as it passes through an object. Bacon depicts the motion of species as a radiation that multiplies species as similar to the production of light from light.[37]

Naturally, based on Bacon's belief in the similarity between light and species, light then serves as a model in optics, as well as in geometry and physics. It is rather difficult to construct an exhaustive list of all the ways in which Bacon treats light as a model for species in his *De Perspectiva*. This text contains extensive discussion and diagrams of the movement of species that are modeled on the movement of light. One example will have to suffice. Bacon uses the reflection of light in mirrors to understand how other objects are seen in mirrors.[38] By doing so, Bacon is able to argue that celestial bodies do not reflect the sun, since if this were the case, the sun would be seen in the stars and moon. He is also able to explain why objects appear differently in a mirror than how they are in reality. Moreover, Bacon criticizes theologians who attempt to answer

---

12 and Roger Bacon, *On Signs*, trans. Thomas S. Maloney (Toronto: Pontifical Institute of Mediaeval Studies, 2013).

[34] Bacon, *On the Multiplication of Species*, 2. 1. 57–60.

[35] Ibid., 4. 3. 60–63.

[36] Bacon, *Appendix* to *On the Multiplication of Species*, pp. 88–94 in David C. Lindberg, *Roger Bacon's Philosophy of Nature*.

[37] Bacon, *On the Multiplication of Species*, 6. 1. 10–27 and 5. 1. 6–7.

[38] Bacon, *Perspectiva*, III, d. 1, c. 1–3 in Lindberg, *Roger Bacon and the Origins of Perspectiva in the Middle Ages*.

questions about the nature and motion of light without a working knowledge of geometry, arguing,

> these matters in no way can be known without what has been said concerning multiplication according to lines, angles, and figures. For the multiplication of light is just like the multiplication of every other species of any agent whatsoever.[39]

Shortly after arguing for the utility of mathematics for other sciences, Bacon continues on to expound on the usefulness of species in understanding the force exerted on the earth by celestial beings through the propagation of species. Upon this foundation, Bacon explains the motion of bodies in the cosmos. Bacon cites light emanating from the sun as the paradigmatic example to comprehend motion through bodies of varying density. The entirety of this discussion is placed within the context of Bacon's analysis of theoretical and practical astrology.[40] The motion of species as modeled upon the action of light had clear importance in elucidating other natural sciences for Bacon.

However light's contribution is not limited to these fields for Bacon. Although he is speaking of the larger Medieval period, Lindberg rightly points toward the connections Bacon saw between light and some other less likely academic fields when he comments:

> If these opinions (on the nature of light and the act of vision) were affiliated with any specific subject-matter, it would have been with cosmology, meteorology, psychology, or metaphysics: cosmology because the principle source of light were celestial bodies; meteorology because of the presence of striking luminous phenomena such as the rainbow and halo in the atmosphere; psychology because of the pre-eminence of sight among the five external senses; and metaphysics because of the ubiquity of light as analogue and metaphor in the works of Neoplatonic metaphysicians. Light also figured in theological discussions, owing to its prominence in the creation account in Genesis, the biblical use of metaphors, and the close connection between theology and metaphysics.[41]

In fact, Bacon does address several of the disciplines Lindberg mentions in regard to light. It would take too long to fully examine all of the ways Bacon views these disciplines, thus, I will only cite a few examples. In addressing theological concerns, he contends that the infusion of grace is

---

[39] Bacon, *Opus maius*, 4, dist. 4, ch. 16.
[40] Ibid., 4. Dist. 2, ch. 1–2.
[41] Lindberg, *Roger Bacon and the Origins of Perspective in the Middle Ages*, pp. xxv–xxvi.

'very clearly illustrated though the multiplication of light' holding that in a saint the infusion of grace is similar to light moving directly and perpendicularly, while in an imperfect person the light is refracted, and, of course, in a sinner the light of grace is driven away or reflected.[42] Regarding meteorology, Bacon attempts to explain the nature of a rainbow, a complex task that required him to consider motion of the observer, light, and the sun, as well as the effect clouds might have on light and the appearance of color in what should be invisible light.[43] In his discussion of mathematics, Bacon himself explains the utility of the study of species:

> every efficient cause acts by its own force which it produces on the matter subject to it, as the light of the sun produces its own force in the air, and this force is light diffused through the whole world from the solar light. This force is called likeness, image, species, and by many other names, and it is produced by substance as well as accident and by spiritual substance as well as corporeal. Substance is more productive of it than accident, and spiritual substance than corporeal. This species causes every action in this world; for it acts on sense, on intellect, and all the matter in the world for the production of things, because one and the same thing is done by a natural agent on whatever it acts [...] But if it acts on the sense and the intellect, it becomes a species, as all know. Accordingly, on the other hand, if it acts on matter it also becomes a species.[44]

---

[42] Bacon, *Opus maius*, 4, dist. 4, ch. 16.

[43] Bacon, *Opus maius*, 6. 2–11. See also David C. Lindberg, 'Roger Bacon's Theory of the Rainbow: Progress of Regress?', in *Studies in the History of Medieval Optics*.

[44] Ibid., 4, dist. 2, ch. 1. Although there appear to be two different sources of species, Bacon does not want to admit that species are spiritual or immaterial. He contends, 'But when they say that species has a spiritual existence in a medium, this use of the word spiritual is not in accordance with its proper and primary signification, from spirit, as we say, that God and angel and soul are spiritual things; because it is plain that the species of corporeal things are not thus spiritual. Therefore of necessity they will have a corporeal existence, because body and soul are opposed without an intermediate.' Bacon continues on to contradict his earlier statement by claiming that species derive from corporeal things, not spiritual. He tries to circumvent the problem by suggesting that when Aristotle and Alhacen described species as spiritual, they actually meant that the species were merely insensible 'since everything truly spiritual, such as God, angels, and the soul, is insensible and falls outside the province of sense, therefore we employ the terms interchangeably and call insensibles "spiritual"'. He goes to great lengths to explain why this must be so, but his argument can be summed as maintaining that species must be similar to its agent and cannot be more noble than the agent. Therefore, the species must be corporeal. Bacon never seems to work out the tension he alludes to when he seems

There are two important points to be drawn from this statement. First, Bacon suggests that there are two sources of species, spiritual and corporeal, that can act on the senses and the intellect. Second, Bacon again explicitly cites light as an example in relation to the movement of both kinds of species. The *lux-lumen* distinction, then, would seem to cut across disciplinary boundaries as a valid explanatory model based on the simple fact that the categories spiritual and corporeal are universally inclusive.

Further, if we consider Bacon's attitude toward two particular sciences, mathematics and optics, which he held in especially high regard and properly view the priority of light within the study of each of these disciplines, the importance of the *lux-lumen* distinction only becomes more apparent. He describes mathematics as 'the gate and key' of all other sciences. Bacon adds,

> The knowledge of this science prepares the mind and elevates it to a certain knowledge of all things, so that if one learns the roots of knowledge placed about it and rightly applies them to the knowledge of the other sciences and matters, he will then be able to know all that follows without error and doubt, easily and effectually.[45]

As such, mathematics becomes an important tool for theology insofar as it can open the other sciences which help one to understand the Bible. Even the doctrine of the Trinity can be shown through geometry.[46] As light is a universal analogy that explains causation for Bacon, so is mathematics a universal hermeneutical skeleton key. Regarding optics, Bacon comments, 'It is possible that some other science may be more useful, but no other science has so much sweetness and beauty of utility. Therefore it is the flower of the whole of philosophy and through it, and not without it, can the other sciences be made known.' It is through optics that we can conduct experiments and learn apart from authority. Optics, for Bacon, also elevates human beings above animals who are solely concerned with things pertaining to touch and taste.[47] It is not a coincidence that Bacon addresses mathematics and optics after laying the

---

to admit that truly spiritual substances can emit species. See Bacon, *Opus maius*, 5. 1, dist. 6, ch. 4, Bacon, *Perspectiva*, I, d. 6, c. 4 in Lindberg, *Roger Bacon and the Origins of Perspectiva in the Middle Ages* and Bacon, *On the Multiplication of Species*, 3. 2–3.

[45] Bacon, *Opus maius*, 4, dist. 1, ch. 1.

[46] Ibid., 4, dist. 4, ch. 16.

[47] Ibid., 5.1, dist. 1, ch. 1.

foundation for study with his apology for languages in the *Opus Maius*, but before moving on to the other great sciences, experimental science and moral philosophy. Mathematics and optics are clearly given high priority as indispensable tools for further study in Bacon's thought. Since the *lux-lumen* distinction lies at the heart of mathematics as it serves as a model for Bacon's understanding the motion of species, as pointed to in the earlier discussion of geometry, and it also has fairly clear ramifications for the study of optics, Bacon's attitude toward these two sciences would seem to confirm the centrality of the *lux-lumen* distinction for all of Bacon's thought.

As the last portion of Lindberg's statement above suggests, light ultimately draws Bacon back to God through his understanding of the relationship between the motion of species and the whole of his cosmology. Bacon's cosmology is based on a universal radiation of force. It has already been mentioned above that Bacon defines a species as the 'first effect of any naturally-acting thing'. This is really only the first step in Bacon's cosmology. Bacon envisions a universe in which agents act on and react to one another. He argues, 'When it is said that the agent receives physical action and that the recipient acts physically, this is not simply a matter of resisting; rather, the recipient transforms and alters the agent.'[48] This chain of causation proceeds nearly *ad infinitum* as Bacon contends that a 'species produced [by an agent] in the first part of the air [or other medium] is not separated from that part, since form cannot be separated from the matter in which it is unless it should be mind; rather, it produces a likeness to itself in the second part of the air, and so on'.[49] John Henry Bridges describes the process more precisely, 'The agent acts on the first part of the body of the patient, and stimulates its latent energy to the generation of species. That part thus transmuted acts on the next part succeeding; and so the action proceeds.'[50]

Bacon's starting point in the discussion of the propagation of universal force is the fact that an infinite multiplication of species moving in radiant fashion can result from a single point, just as light radiates in many directions from a single point.[51] Light and all other visible agents

---

[48] Bacon, *On the Multiplication of Species*, 1. 4.37–40.
[49] Bacon, *Opus maius*, 5. 1, dist. 9, ch. 4.
[50] John Henry Bridges, *The Life and Work of Roger Bacon: An Introduction to the* Opus maius (London: Williams & Norgate, 1914; repr. Merrick, NY: Richwood Pub. Co., 1976), p. 97.
[51] Bacon, *On the Multiplication of Species*, 2. 1. 25–30.

are merely particular manifestations of this universal force. In this manner, Bacon is able to trace his way back to God. In the last portion of the *Opus maius*, Bacon tries to prove the existence of God using the proof of God as the prime mover. He observes that, 'Causes do not go back endlessly, since they cannot be infinite in number nor can they be conceived of [...] There is not therefore a cause preceding a cause endlessly. Therefore we must stop at some first cause, which does not have a cause antecedent to it.' He moves from this point to show that this also means that God is eternally existent and infinitely powerful.[52] Thus, Bacon is able to use light to return to its source in the divinity in the same way that *lumen* can be traced back to *lux*.

It should be clear that Bacon's version of the metaphysics of light binds together his massive endeavor. In turn, Bacon's metaphysics of light is unified by the example of the *lux-lumen* distinction. He begins with the conviction that all knowledge stems from divine illumination. This knowledge is imparted in a fashion similar to the way *lux* is distinguished from *lumen*. This process of separation also serves as a model for universal motion as Bacon explicitly cites light as the exemplar for the motion of all species as well as a model for creation through the dissemination of species. As originating from God, creation bears the resemblance of the divine, just as a species resembles its agent. This becomes clearer when Bacon begins to argue for the existence of God by viewing God as the prime mover within a cosmology of force. 'Bacon did not conceive light merely as the instrument of vision,' Lindberg correctly observes, 'he regarded it also as the key to an understanding of the inner workings of nature, a privileged (because of its visibility) case of universal causal agency and, consequently, an opportunity for the most intimate

---

[52] Bacon, *Opus maius*, 7. 4. This is an important point in which Bacon seems to split the difference between some Aristotelians and Christian thinkers such as William of Auvergne. In his *De universo*, William writes against some philosophers who believe that God created through inferior intelligences. Bacon affirms that God alone created, although he admits that God could have used a medium to create out of his goodness. Ultimately, though, these inferior intelligences could not have created on their own. Stewart Easton comments, 'Bacon always insists that the influx of power from the primal cause is the deciding factor in creation.' Inferior intelligences simply do not have the power to create without an infusion from God. See Easton, *Roger Bacon and His Search for a Universal Science*, 50–52 and Roger Bacon, *Questiones supra librum de causis* in *Opera hactenus inedita Rogeri Baconi*, fasc. 12, 105. 26–107. 23 (see n. 15 above).

inspection of natural causes at work.'[53] Bacon's attempt to mathematize nature perhaps predisposed him toward an understanding of the universe based on the *lux-lumen* distinction insofar as it provided a natural explanatory causal model with nearly universal utility in that light is easily subject to mathematic quantification and measurement, but this is certainly subject to debate.[54] Although there are areas in which a connection with light is less obvious—languages, for example—it should be clear that this erstwhile Oxonian's conceptualization of light founded upon the *lux-lumen* distinction unites his intellectual enterprise.[55]

---

[53] Lindberg, 'Light, Vision, and the Universal Emanation of Force', in *Roger Bacon and the Sciences*, ed. Hackett, p. 243.

[54] See David C. Lindberg, 'On the Applicability of Mathematics to Nature', in *Studies in the History of Medieval Optics*.

[55] Thanks are due to the Graduate and Professional Student Association and the Department of Religious Studies at the University of Pennsylvania for monetary contributions to cover travel expenses for this conference.

'THE LIGHT SO IN MY FACE / BIGAN TO SMYTE':
ILLUMINATING LYDGATE'S *TEMPLE OF GLAS*

William T. Rossiter

The *Temple of Glas* poses a critical problem by means of its apparently static simplicity on the one hand, and its claims to allegory on the other. At its close, the narrator promises '[f]orto expoune my forseid visioun, / And tel in plein the significaunce' (ll. 1389–90) in an as yet unspecified future work.[1] Yet the poem is ostensibly plain already, and herein lies the root of the varied responses to it, a brief outline of which I will provide following an equally brief summary of its action, or lack of it, as the case may be.[2]

The poem begins in darkness with the voice of the melancholic narrator, who bears all the characteristics of someone suffering from *aegritudo amoris*, or lovesickness, although this is not made explicit.[3] The narrator proceeds to relate a dream he had 'this othir nyght' (l. 3), in which he found himself in a temple of glass. Within the temple, which is

---

[1] All citations of the *Temple of Glas* are taken from *Lydgate's Temple of Glas*, ed. J. Schick, EETS, e.s. 60 (London: Kegan Paul, Trench, Trübner, 1891). Thorn and yogh have been modernized. See also Julia Boffey, ed., *Fifteenth-Century English Dream Visions: An Anthology* (Oxford: Oxford University Press, 2003), pp. 15–89. Schick draws on Oxford: Bodleian, MS. Tanner 346, whereas Boffey uses BL, Add. MS 16165 (=S) as her primary source. For a list of the manuscripts, see Boffey, pp. 18–19. Also in the Tanner manuscript are Chaucer's *Book of the Duchess* and *The Parliament of Fowls*, both of which are sources for Lydgate's poem. See *Manuscript Tanner 346: A Facsimile*, intro. Pamela Robinson, The Facsimile Series of the Works of Geoffrey Chaucer, vol. 1 (Norman, OK: Pilgrim, 1980), fol. 76ʳ–97ʳ.

[2] *Plein* here, however, tends towards 'completely' rather than 'clearly' (*MED*). Both usages appear in works by Chaucer upon which Lydgate draws in the poem. The first usage corresponds with 'thei kan nought pleynly understonde' (*Tr*, II. 272); the second usage corresponds with '[s]peketh so pleyn at this time, we yow preye, / That we may understonde what ye seye' (*CT*, IV. 19–20). All citations of Chaucer's works are taken from *The Riverside Chaucer*, ed. Larry D. Benson et al., 3rd edn (Oxford: Oxford University Press, 1988).

[3] Lovesickness was addressed by treatises such as Bernard of Gordon's *Lilium medicinae* (c. 1285), and was considered fatal in extreme cases. See J. L. Lowes, 'The Loveres Malady of Hereos', *Modern Philology*, 11 (1914), 491–546.

dedicated to Venus, are numerous images of unfortunate lovers, in addition to various plaintiffs pleading their amorous cases before the goddess.[4] Amongst these one lady in particular shines out—literally—as she tells of her misfortune. The dreamer also describes a knight who complains of unrequited, unspoken love in typical courtly terms. In the final part of the poem, the dreamer describes how the knight professes his love to the lady before Venus, who binds them together in quasi-matrimonial terms and exhorts them to patience. The temple then resounds with an harmonious hymn to the goddess, which wakes the dreamer.

Commentary upon the poem may be divided into two interrelated themes, namely de-allegorization and binary opposition. In relation to the former, a number of commentators have identified a move towards greater realism which, whilst it enables focus upon the poem's concrete historical moment, also gives rise to the criticism that Lydgate has stripped away the allegorical complexity attendant upon the Chaucerian dream-vision.[5] A. C. Spearing, for example, argues that 'Lydgate is attracted by picturesque details [...] [b]ut he is interested in such things simply for their own sake, as a magpie is attracted by anything shiny'.[6] The secondary theme, that of binary opposition, is predicated upon the Boethian doctrine of contraries, whereby one thing may only be understood through its antithesis:

> And of thise thinges, certes, everiche of hem is declared and schewed by other. For so as good and yvel ben two contraries, yif so be that good be stedfast, thanne scheweth the feblesse of yvel al opynly; and yif thow knowe

---

[4] James Simpson discusses the political valences of such courtly *plaint* in his *Reform and Cultural Revolution*, The Oxford English Literary History, volume 2: 1350–1547 (Oxford: Oxford University Press, 2002), pp. 121–90. Indeed, J. Allan Mitchell has recently argued that the *Temple of Glass* allegorizes Katherine of Valois' covert (yet well-known) liaison with Owen ap Tudor following the death of Henry V in 1422, a reading which necessarily reinforces the later dating of the poem (cf. n. 18 below). See 'Queen Katherine and the Secret of Lydgate's Temple of Glass', *Medium Ævum*, 77 (2008), 53–76.

[5] C. S. Lewis, *The Allegory of Love: A Study in Medieval Tradition* (Oxford: Oxford University Press, 1936), pp. 239–43.

[6] A. C. Spearing, *Medieval Dream-Poetry* (Cambridge: Cambridge University Press, 1976), p. 173.

clerly the freelnesse of yvel, the stedfastnesse of good is knowen. (*Boece* IV, pr. 2. 10–12)[7]

The doctrine of contraries is encapsulated in the poem by Lydgate's comment that 'white is whitter, if it be set bi blak' (l. 1250), and developed into the wider diagnosis that 'Lydgate's mind, like a computer, operates thus on a binary system [...] so that every human situation becomes a dilemma, and every shade of feeling is resolved into a straight antithesis'.[8]

The Boethian interpretation of the poem is important both in terms of linking the dream frame to the dream proper by means of a structural reflection which allows one to question the assumption that Lydgate has simply written an evacuated allegory.[9] For example, there are various verbal and imagistic echoes bounced back and forth between the dreamer and the knight, between the lady and Venus, between the knight and the lady. More importantly for the present study, the doctrine of contraries inevitably emphasizes the careful balance between light and dark which Lydgate maintains throughout, to the extent that *chiaroscuro* itself becomes the poem's organizing principle. Indeed, Judith Davidoff argues that

> the *Temple of Glas* is a poem primarily about 'light' and that Lydgate used his dream frame for two ends: first, to raise the issue of how a poet can use poetic language to shed light on his own darkness and, second, to teach his audience how to 'read' his poem and by extension to apply the light / darkness theme to themselves.[10]

I agree entirely with the assertion of light's centrality to the poem; even those commentators who are less than smitten use light imagery to

---

[7] See John Lydgate, *Poems*, ed. John Norton-Smith (Oxford: Clarendon Press, 1966), p. 177, and *Tr*, I. 638–44.

[8] Derek Pearsall, *John Lydgate* (London: Routledge and Kegan Paul, 1970), pp. 113–14.

[9] See Janet Wilson, 'Poet and Patron in Early Fifteenth-century England: John Lydgate's *Temple of Glas*', *Parergon*, 11 (1975), 25–32; Judith M. Davidoff, 'The Audience Illuminated, or New Light Shed on the Dream Frame of Lydgate's *Temple of Glas*', *Studies in the Age Chaucer*, 5 (1983), 103–25; and Anna Torti, *The Glass of Form: Mirroring Structures from Chaucer to Skelton* (Cambridge: Brewer, 1991), pp. 67–86. See also Lois Ebin, *Illuminator, Makar, Vates: Visions of Poetry in the Fifteenth Century* (Lincoln, NE: University of Nebraska Press, 1988), pp. 19–24.

[10] Davidoff, 'The Audience Illuminated', p. 104.

describe it, as Spearing's magpie simile shows. However, whereas Davidoff discusses the interplay between illumination and darkness from a primarily metaphorical perspective – which is understandable given the problem of the *Temple*'s allegoresis – I intend to discuss how Lydgate's poem relates to the various contemporary discourses upon light, and the concomitant discourses of vision and reflection. In particular I will focus upon the roles of light within optics, theology, and of course poetry, and how these might be seen to inform the poem's structure and its potential for meaning thereby, particularly in relation to the opening *chiaroscuro* and the images of the lady and the goddess. What will become apparent is Lydgate's balancing of discursive multiplicity (optical, theological, poetical) with a single theme (light). The effect of this balancing act, which is characteristic of Lydgatean *copia*, is hermeneutic open-endedness. The proliferation of illuminating discourses, rather than casting light upon the poem's 'significaunce', tends to leave the reader in the dark; a position which s/he shares with the dreamer at the close of the poem.

The poem begins with a complex interplay of light and dark imagery, providing the basis upon which the rest of the poem may build:

> For thought, constreint, and greuous heuines,
> For pensifhede, and for heigh distres,
> To bed I went nov this othir nyght,
> Whan that Lucina with hir pale light
> Was Ioyned last with Phebus in aquarie,
> Amyd decembre, when of Ianuarie
> Ther be kalendes of the nwe yere,
> And derk Diane, ihorned, nothing clere,
> Had [hid] hir bemys vndir a mysty cloude:
> Within my bed for sore I gan me shroude,
> Al desolate for constreint of my wo,
> The long[e] nyght waloing to and fro,
> Til at[te] last, er I gan taken kepe,
> Me did oppresse a sodein dedeli slepe,
> With-in the which me thought[e] that I was
> Rauysshid in spirit in [a] temple of glas—
> I nyst[e] how, ful fer in wildirnes—
> That foundid was, as bi lik[ly]nesse,
> Not opon stele, but on a craggy roche,
> Like ise Ifrore. (ll. 1–20)

The poem opens upon darkness which reflects the narrator's 'thought, constreint, and greuous heuines' (l. 1), and as such is literally a dark conceit. The moon appears in two forms, as 'Lucina with hir pale light

[...] And derk Diane, ihorned, nothing clere', who hid 'hir bemys vndir a mysty cloude' (ll. 8–9). Lucina's pale light is an absent presence therefore, included as a characteristic epithet which is nevertheless negated. However, after being oppressed by his 'sodein dedeli slepe' the narrator finds himself '[r]auysshid in spirit in [a] temple of glas', which reflects the light from the sun into his face and blinds him until cloud-cover re-enables vision:

>   And as I did approche,
> Again the sonne that shone, me thought, so clere
> As eny cristal, and euer nere and nere
> As I gan neigh this grisli, dredful place,
> I wex astonyed: the light so in my face
> Bigan to smyte, so persing euer in one
> On euere part, where that I gan gone,
> That I ne myght nothing, as I would,
> Abouten me considre and bihold,
> The wondre estres, for brightnes of the sonne;
> Til at[te] last certein skyes donne,
> With wind Ichaced, haue her cours Iwent
> To-fore the stremes of Titan and Iblent,
> So that I myght, with-in and with-oute,
> Where so I walk, biholden me aboute,
> Forto report the fasoun and manere
> Of al this place, that was circulere
> In compaswise, Round bentaile wrought. (ll. 20–37)

Whilst one can interpret this movement from darkness to light and back again in accordance with the Boethian doctrine of contraries, there seems to be more going on here in terms of illumination's necessity for vision.[11]

---

[11] The movement from blindness to sight is suggestive (for the modern reader at least) of Plato's cave, as detailed in the *Republic*, 514a–518b. Interestingly, Lydgate's patron Humphrey, Duke of Gloucester, also patronized a translation of the *Republic* by Pier Candido Decembrio, although this was in the 1430s, long after the usual dates given for the *Temple* (see below). See Pearsall, p. 225. On Humphrey's humanist credentials see Roberto Weiss, *Humanism in England during the Fifteenth Century*, 4th edn, ed. D. Rundle and A. J. Lappin (Oxford: Society for the Study of Medieval Languages and Literature, 2010), Chapter 3 (available online at http://mediumaevum.modhist.ox.ac.uk/monographs_weiss.shtml, and due for print publication in 2015), and more recently Alessandra Petrina, *Cultural Politics in Fifteenth-Century England: The Case of Humphrey, Duke of Gloucester* (Leiden: Brill, 2004), and Daniel Wakelin, *Humanism, Reading, and English Literature 1430–1530* (Oxford: Oxford University Press, 2007), pp. 23–61. See also David Rundle, ed.,

Aside from the neat reflection of the cloud-obscured moon in the dream frame, which disables vision, by the cloud-covered sun in the dream proper, which enables it, the movement between extremes is suggestive of the Aristotelian conception of vision which, according to David Lindberg, had become the predominant model by the early fifteenth century, when the poem is thought to have been written.[12] This did not mean that other models became automatically obsolete. On the contrary, the late medieval tendency towards encyclopaedism and synthesis allowed for the fusion of different modes; and Lydgate was, if nothing else, indicative of his age, and very much given to encyclopaedism. Aristotle's description in *De anima* of how light enables vision by actualizing the interpositive transparent medium certainly appears to correspond with Lydgate's apparently intromissive and sudden blinding, his account of how 'the light so *in* my face / Bigan to smyte' (emphasis added):

> There is, accordingly, something transparent. By transparent I mean that which is, indeed, visible, yet not of itself, or absolutely, but by virtue of concomitant colour. Air and water and many solids are such. [...] Light is the act of this transparency, as such: but in potency this [transparency] is also darkness. Now, light is a kind of colour of the transparent, in so far as this is actualised by fire or something similar to the celestial body; which contains indeed something of one and the same nature as fire. (II, Lecture XIV, 404–05 [418b4-13])

This description was in turn elucidated by Aquinas's commentary:

> To begin with, therefore, he says that if colour is that which of its nature affects the transparent, the latter must be, and in fact is, that which has no intrinsic colour to make it visible of itself, but is receptive of colour from without in a way which renders it somehow visible. Examples of the transparent are air and water and many solid bodies, such as certain jewels and glass. [...] Of itself the transparent is in potency to both light and darkness (the latter being a privation of light) as primary matter is in potency both to form and the privation of form. Now light is to the transparent as colour is to a body of definite dimensions: each is the act and form of that which receives it. And on this account he says that light is the colour, as it were, of the transparent, in virtue of which the transparent is made actually

---

Humanism in Fifteenth-Century Europe (Oxford: Society for the Study of Medieval Languages and Literature, 2012).

[12] D. Lindberg, *Theories of Vision from Al-Kindi to Kepler* (Chicago: University of Chicago Press, 1976), p. 116.

so by some light–giving body, such as fire, or anything else of that kind, or by a celestial body. (II. 7. 404–05)[13]

The dreamer is thus the comparatively passive, Aristotelian recipient, rather than the extramissive Platonic agent.[14] The transparent in Lydgate's poem is actualised by the celestial body, that is, the 'brightnes of the sonne'. The Temple itself, considering its material, may by extension be read as symbolic of the Aristotelian transparent. Aquinas specifically refers to 'certain jewels and glass' as examples of the transparent, and indeed, Lydgate describes how the sun shone 'so clere / As eny crystal [...] so persing ever in one / On evere part'. This description may even be interpreted as referring to the punctiform nature of light as it is described in the Latin translation of Ptolemy's *Optica*: 'every visual ray reaches a single point [*unius puncti*] which is proper to it'.[15] Ptolemy was an influence upon the Arabic optician and physicist ibn al-Haytham, better known in the West as Alhazen, whose work in turn influenced Roger Bacon. Bacon's theory of vision is predicated upon species (rays) issuing from all points on the viewed object to all points on the eye's surface, forming a cone of vision the apex of which is the centre of the eye. Interestingly, Bacon's language shifts between the point and the whole:

---

[13] *Commentary on Aristotle's De anima*, trans. Kenelm Foster and Silvester Humphries, intro. Ralph McInerny (Notre Dame, IN: Dumb Ox Books, 1994), pp. 128–30. For the Latin text see Sancti Thomas De Aquino, *Opera Omnia*, iussu Leonis XIII P.M. edita, Tomus XLV, *Sentencia libri De anima*, cura et studio Fratrum Praedicatorum (Rome & Paris: Commissio Leonina & J. Vrin, 1984), pp. 123–25.

[14] As Robert Nelson argues in his introductory essay to R. S. Nelson ed., *Visuality Before and Beyond the Renaissance: Seeing as Others Saw* (Cambridge: Cambridge University Press, 2000): 'These theories have implications for the relation of the observer to the observed. Extramission implies that vision is active and motivated and that it causes subject and object to have direct physical contact. Intromission is more passive, but all vision in antiquity and the Middle Ages was comparatively active' (pp. 4–5). See also the essay by Michael Camille in the same volume 'Before the Gaze: The Internal Senses and Late Medieval Practices of Seeing', pp. 197–223, esp. pp. 204–08. The Platonic extramissive model was available to the late Middle Ages via Chalcidius' commentary on the *Timaeus*, although the Aristotelian and Platonic models are not as distinct as they might appear, as the thirteenth-century optical syntheses show.

[15] Cited in Simon A. Gilson, *Medieval Optics and Theories of Light in the Works of Dante* (Lewiston, NY: Mellen, 2000), p. 12. See also Albert Lejeune, ed., *L'optique de Claude Ptolémée dans la version latine d'après l'arabe de l'émir Eugène de Sicile*, new edn (Leiden: Brill, 1989), pp. 20–21.

Although to every point of the eye and cornea comes the apex of one pyramid [originating] from the whole object, and the species of all parts [of the object] are there mixed, nevertheless to one point of the eye or cornea and the aperture of the uvea [i.e. the pupil] comes a species perpendicularly from only one point of the visible object, although to the same point come an infinity of species inclined at unequal angles.[16]

Lydgate's description of the light 'persing ever in one / On evere part' may thus be read as being informed by contemporary optics, echoing as it does Ptolemy's 'unius puncti' and Bacon's explanation of the emanation of species from 'the whole object' both to 'every point of the eye' and 'to one point of the eye'.

Lydgate's encyclopaedism, furthermore, and the fact that optics had already fused various discourses, also allows for a possible Platonic reading, again based upon the Temple's material. Whilst a synthesized form of Aristotelian thought might have assumed the optical hegemony, arguably the most readily available discourse upon reflection—outside of observations by Seneca and Pliny—stemmed from Chalcidius' partial translation of the *Timaeus* and again from Ptolemy's *Optica*, which as we know influenced Alhazen, Bacon and those who followed them.[17] Yet aside from the basic fact of reflection (*intuitio*) and refraction (*detuitio*), it is questionable as to how Lydgate's knowledge of catoptrics could be said to inform the poem's imagery, save to note that the temple provides a surface by which the sunlight reaches the dreamer; and as we see in the excerpt from *De anima*, the presence of celestial light and transparent materials already figure in the Aristotelian model. In other words, the move from darkness to light would appear illustrative of the late medieval Aristotelian model of intromission, whilst it remains possible that Lydgate's concept of reflection might have been informed by the relevant

---

[16] Translated in Lindberg, p. 109 (see n. 12 above). See also John H. Bridges, ed., *The Opus maius of Roger Bacon*, 3 vols (London: Williams and Norgate, 1900), II, 37–38.

[17] Seneca's discussion of mirrors is linked to his discussion of the rainbow. See *Naturales Quaestiones I*, trans. T. H. Corcoran, Seneca in Ten Volumes (Cambridge, MA: Harvard University Press, 1971), VII, 42–55 (44–45). See also Pliny, *Natural History*, trans. H. Rackham, 10 vols (Cambridge, MA: Harvard University Press, 1952; repr. 1961), IX, 96–99. For an account of reflection as treated by Aristotle see Carl B. Boyer, 'Aristotelian References to the Law of Reflection', *Isis*, 36.2 (1946), 92–95. In the discussion of sound and echoes which immediately follows the discussion of light and vision in *De anima*, Aristotle notes that light 'is always reflected' (p. 278).

sections in the Chalcidian *Timaeus* or Ptolemy's *Optica* and their commentaries—all of which Lydgate could have had access to either via the impressive library at Bury St Edmunds, where he was a monk, or at Oxford, where Lydgate studied at Gloucester College (*c*.1406–1408).[18] In any case I am not nominating Lydgate as a close reader of the commentaries and the *perspectivae*; what have been discussed are the basics, which were well-established by the early fifteenth century.[19]

It may be argued that the emphasis placed upon this act of seeing, which corresponds with the physical process, contrasts with the idea of being 'in spirit'. Yet it is possible to envisage the subject as embodied and responsive to physical stimuli within a vision, as we find so often in dream poetry. One thinks, for example, of Chaucer's narrator in the *Book of the Duchess*, waking 'al naked' into the dream, after being 'affrayed' by the 'noyse and swetnesse' of birdsong (*BD*, 293–97). As Renoir argues, '[n]ot only does Lydgate know how to write a lover's complaint, but he also knows how to set the physical background' (p. 24). It is not a

---

[18] The possibility that Lydgate was influenced by optical discourse whilst at Oxford refutes the dating of the poem as being written *c*.1400–3, the date given by Schick in his introduction to *TG* (p. c). Wilson (p. 26) also argues that 'the earlier date of 1403 proposed by Schick and Schirmer is the more likely one', whilst Alan Renoir suggests a possible later date—1420—due to the reference to the Paston family motto in line 310 (William Paston married Agnes Bury in 1420): A. Renoir *The Poetry of John Lydgate* (London: Routledge and Kegan Paul, 1967), p. 50. Boffey, *English Dream Visions,* p. 16 likewise claims there is 'no real support for this view [that is, the earlier dating] and indeed the outlines of Lydgate's career suggest that his middle years were the period of most intense activity on behalf of the secular patrons and audiences who might have found *TG* to their taste'. Pearsall also claims that the courtly poems were probably written in the 1420s–early 1430s: D. Pearsall, *John Lydgate (1371–1449): A Bio-bibliography*, ELS Monograph Series, 71 (Victoria, BC: University of Victoria, 1997), pp. 14, 31. Pearsall's estimate accords with that of Mitchell (p. 63), who claims between 1427 and 1432 for the poem's composition; indeed, the later dating is crucial to his allegoresis (see n. 4 above). On Lydgate's Oxford, see J. I. Catto and R. Evans, eds, *The History of the University of Oxford, Volume II: Late Medieval Oxford* (Oxford: Oxford University Press, 1992) and A. B. Emden, *A Biographical Register of the University of Oxford to A. D. 1500*, 3 vols (Oxford: Oxford University Press, 1957–59), II, col. 1185–86. The Norman tower at Bury might have fired Lydgate's imagination, as Walter Schirmer describes how '[l]ight poured in through the twelve brightly-coloured stained glass windows': W. Schirmer, *John Lydgate: A Study in the Culture of the XVth Century*, trans. Ann E. Keep (London: Methuen, 1961), p. 12.

[19] See Lindberg, pp. 120–22.

contradiction to say that the dream poem relates the experience of the embodied consciousness whilst disembodied.

Aside from optics, the discourse of illumination also features in late medieval theology, and this also may have some bearing upon the poem's *significaunce*. That is, if neo-Aristotelian optical discourse explains how light reaches and affects the dreamer, the theological aspect might help clarify how it is to be interpreted. For example, in the commentary upon *De anima*, Aquinas, whilst denying that sensible light is 'spiritual in nature', concedes that 'if anyone should say that there is a spiritual 'light' other than the light that is sense-perceived, we need not quarrel with him [...] For there is no reason why quite different things should not have the same name' (II. 7. 415–16).[20]

However, the most influential figure in the theology of vision, as Cynthia Hahn has noted, is Augustine.[21] Augustine affirms that 'God is the archetypal light, and sensible light is the imitation' (cited in Lindberg, p. 96), drawing a comparison between God and Plato's perfect form of the Good. Robert Grosseteste would later refer to this analogy in *De veritate*, wherein he posits that 'just as infirm corporeal eyes do not see coloured bodies unless they are illuminated by the sun [...] so the infirm eyes of the mind do not perceive truths themselves except in the light of the supreme truth' (ibid.). Furthermore, in *De Genesi ad litteram* Augustine distinguishes between three forms of vision: corporeal vision, spiritual vision—such as occurs in dreams and the imagination—and finally intellectual vision, which he illustrates via II Corinthians 12:

> Scio hominem in Christo ante annos quatuordecim, sive in corpore nescio, sive extra corpus nescio, Deus scit, raptum hujusmodi usque ad tertium cælum [...] raptus est in paradisum
>
> I knew a man in Christ above fourteen years ago, (whether in the body, or out of the body, I cannot tell: God knoweth;) such an one caught up to the third heaven [...] he was caught up into paradise (II Corinthians 12. 2–4)

---

[20] Aquinas nevertheless explains that the reason 'why we employ "light" and other words referring to vision in matters concerning the intellect is that the sense of sight has a special dignity; it is more spiritual and more subtle than any other sense' (*De anima*, II. Lect. XIV, 417), although it is not spiritual *per se*.

[21] On the influence of Augustine as 'the foremost "theologian" of vision' see Cynthia Hahn, '*Visio Dei*: Changes in Medieval Visuality', in *Visuality Before and Beyond the Renaissance*, pp. 169–96 (pp. 169–74). Hahn cites *De Genesi ad litteram* as the most important medieval text on vision, although in terms of light one might also see book XI of *De trinitate*. See also Lindberg, pp. 95–96.

Paul's language appears to be echoed by Lydgate's explanation of how he came to the temple, that is, of being '[r]auysshid in spirit'. In fact, there are three points of interest here. The first is the linguistic correspondence which becomes more apparent when we compare the Vulgate's account of the 'hominem in Christo [...] raptum [...] usque tertium caelum'.[22] 'Rauysshid' corresponds with 'raptum' clearly, whilst 'in spirit' may be understood as being periphrastic of 'extra corpus'.[23] The second point concerns the *tertium caelum*, or third heaven. Scholars have recently claimed that Paul's phrase draws on Jewish mysticism, yet for a late medieval poet Paul's third heaven may be seen to correspond with the sphere of Venus as it figures in Ptolemaic cosmology.[24] The third point concerns the intertextual relations of the Pauline passage. In particular, the account of the vision has been read as autobiographical, and as informing the Damascene conversion, which of course involved Paul's being blinded by a bright light: 'et subito circumfulsit eum lux de caelo [...] surrexit autem Saulus de terra apertisque oculis nihil videbat' ('And suddenly a light from heaven shined round about him [...] And Saul arose from the ground: and when his eyes were opened, he saw nothing', Acts 9: 3, 8).

There are thus a number of theological interpretations made possible by the narrator's experience of the blinding light. It may be read as an Augustinian spiritual vision—as it is, after all, a dream poem—although the description of being '[r]auysshid in spirit' also echoes the Pauline account Augustine uses to illustrate the intellectual vision's glimpse of beatific knowledge. In which case Venus might also be seen to represent Christian divinity; it was not uncommon for courtly poetry and allegory

---

[22] Larry Scanlon has also argued for the correspondence between Lydgate's phrasing and II Corinthians 12. 2–4 here. See 'Lydgate's Poetics: Laureation and Domesticity in the *Temple of Glass*', in *John Lydgate: Poetry, Culture, and Lancastrian England*, ed. Larry Scanlon and James Simpson (Notre Dame, IN: University of Notre Dame Press, 2006), pp. 61–97 (pp. 74–75). On the link between 2 Cor. 12 and the road to Damascus see Chrysostom's homily XXVI on the Epistles of St Paul to the Corinthians.

[23] The qualifier 'in spirit' also recalls Revelation 4. 2: 'Et statim fui in spiritu' ('And immediately I was in the spirit').

[24] See C. R. A. Morray-Jones, 'Paradise Revisited (2 Cor. 12:1–12): The Jewish Mystical Background of Paul's Apostolate. Part 2: Paul's Heavenly Ascent and Its Significance', *Harvard Theological Review*, 86 (1993), 265–92. See also William Baird, 'Visions, Revelation, and Ministry: Reflections on 2 Cor 12:1–5 and Gal 1:11–17', *Journal of Biblical Literature*, 104 (1985), 651–62.

to conflate Pagan and Christian Love in its divine or perfect form.[25] As Anna Torti argues, in the poem 'Venus becomes a substitute for both God and the Establishment, in that she concentrates in herself and synthesizes the crystallizations of such social qualities as constancy, faithfulness, chastity, and all the stereotypes of Lydgate's society' (p. 77).[26] There also remains the possibility that the blinding light in some way represents a spiritual conversion from the darkness of sin to the light of truth, in which case one is forced to ask what the narrator's or dreamer's sin might be. The answer could be sloth, or more properly *acedia*. As Siegfried Wenzel has noted, *acedia* was not so much akin to our modern conception of sloth as laziness as it was to *melancholia*. The symptoms listed in the poem's opening lines: 'thought, constreint, and greuous heuines [...] pensifhede, and [...] heigh distres' certainly correspond with Wenzel's account of sloth as being 'grief, sorrow, depression, or (to use the Latin equivalent) *tristitia*'.[27] It is worth recalling that *aegritudo amoris* is not specified by the dreamer, and that *acedia*, also known as the noon-day demon, was a sin to which monks were particularly susceptible.[28] Furthermore, if the dreamer is to be enlightened by the vision of the dream, as Davidoff convincingly argues, then it is perhaps worth noting Venus's commendation of the Lady, which declares that 'trouthe [...] feithful menyng, & the Innocence, / That planted bene, withouten eny slouthe, / In your persone, deuoide of al offence' (ll. 377–80), although 'slouthe' here may also hint at its modern meaning (cf. *MED, slouth(e* (n.)). Much of this is, of necessity,

[25] See for example the *prohemium* to book III of *Troilus and Criseyde*, which draws on *Filostrato* (III, 74–79), which in turn draws on Boethius (*De cons.*, 2 metr. 8). The dream in the *Temple* closes with Venus conducting what is effectively a Christian marriage rite. I discussed Chaucer's adaptation of *Filostrato* at length in chapters 2 and 3 of my *Chaucer and Petrarch* (Woodbridge and Rochester, NY: Boydell and Brewer, 2010).

[26] '[S]tereotypes' is perhaps an unfortunate choice of word. And whilst I agree with Torti in relation to Venus's embodiment of orthodoxy, I disagree with the argument that 'because of her iconographic association with the mirror (which by the fifteenth century was made of glass), she is also a symbol of vanity, instability, and transitoriness' (p. 69), at least in this poem.

[27] Siegfried Wenzel, *The Sin of Sloth: Acedia in Medieval Thought and Literature* (Chapel Hill: University of North Carolina Press, 1960; repr. 1967), p. 159.

[28] See Wenzel, p. 5, and Morton W. Bloomfield, *The Seven Deadly Sins: An Introduction to the History of a Religious Concept, with Special Reference to Medieval English Literature* (Michigan: State University Press, 1952; repr. 1967), p. 356 n. 24.

conjectural, as Lydgate fails to provide the poem's 'plein [...] significaunce', although one can be sure that he was conscious of light's symbolical plenitude as it pertained to spiritual illumination.

For the remainder of this study, however, focus will be upon the poetic discourse of illumination. This poetic illumination is predicated on the figures of Venus and the lady. There are two perspectives upon the lady, that of the dreamer and that of the knight; the former details her immediate visual effect, whilst the latter provides a retrospective account of how her image entered through the eye and became imprinted upon the heart. The dreamer compares her beauty to the sun's rays, which of course he has recent experience of:

> ther knelid a ladi in my syght
> Tofore the goddes, which right as the sonne
> Passeth the sterres & doth hir stremes donne,
> And Lucifer, to voide the nyghtes sorow,
> In clerenes passeth erli bi the morow,
> [...]
> Right so this ladi with hir goodli eiye,
> And with the stremes of hir loke so bright,
> Surmounteth al thurugh beaute in my sighte:
> [...]
> Whos sonnyssh here, brighter than gold were,
> Lich Phebus bemys shynyng in his spere—
> [...]
> For in goode faith, thurugh hir heigh presence
> The tempil was enlumynd enviroun
> [...]
> An exemplarie, & mirrour eke was she
> Of secrenes, of trouth, of faythfulnes (ll. 250–95)

The lady appears to embody both the light and the medium of light as described in the poem's opening: she illuminates the temple yet is herself a means of reflection as she is a mirror of truth; and as we know light imagery was used as a means of analogizing the refracted experience of divine truth in the Augustinian theology of vision. The dreamer's reference to 'the stremes of hir loke so bright' is particularly interesting as it would seem to suggest the kind of visual species which Ockham had refuted, whereby the object made visible through illumination transmits a duplicated image through the transparent medium to the eye of the observer.[29] The potential for confusion lies in the inclusion of 'her goodlie eiye', which might render 'hir loke' as her gaze, making the lady the subject as opposed to the object of vision, and perhaps suggesting a return to Platonic extramission, with its conception of interior fire which streams out from the eye.

We might shed some light on Lydgate's sequence of visual events by comparing it with the process followed by Dante both in the *Commedia* and in his early lyrics. Although there is no evidence that Lydgate knew Dante's work directly—outside of what he referred to as 'Dante in Inglissh' (*Fall of Princes*, I. 303), which most commentators assume is a reference to the *House of Fame*—he nevertheless shares with Dante not only an emphasis upon the beloved's blinding beauty, but also what Simon Gilson refers to as the 'motif of dazzling and visual tempering', which effects a 'reciprocity between blinding and recovery of greater vision' (p. 85).[30] Gilson describes how, following the Dante-*personaggio*'s sensual overwhelming at the sight of Beatrice in *Purgatorio* XXXI, the following canto 'opens with Dante's eyes feasting upon his lady to the exclusion of his other senses' (p. 84). In the same way, following his blinding, Lydgate's narrator gives himself over to a visual banquet:

> I fond a wiket, and entrid in as fast
> Into the temple, and myn eiyen cast
> On euere side, now lowe & eft aloft.
> And right anone, as I gan walken soft,
> If I the soth aright report[e] shal,
> I saughe depeynt opon euere wal,
> From est to west, ful many a faire Image
> Of sondri louers, lich as thei were of age
> I-sette in ordre, aftir thei were trwe,
> With lifli colours wondir fressh of hwe. (ll. 39–48)

Also, the Dantean process which Gilson identifies further illuminates Lydgate's poem by means of its provenance. Following his discussion of twelfth- and thirteenth-century optical developments, Gilson points out that whilst 'it is extremely difficult to establish whether Dante knew these

---

[29] Yet it is evidently the same kind of stream as those included in the analogy of the lady with the sun: 'riȝt as þe sonne / Passeþ þe sterres & doþ hir stremes donne'. On Ockham's refutation of species via his concepts of intuitive and abstractive cognition (as discussed in the *Commentary on the Sentences*) see Lindberg, pp. 140–42.

[30] Henry Bergen, ed., *Lydgate's Fall of Princes*, 4 vols, EETS, e.s. 121–24 (London: Oxford University Press, 1924–27, repr. 1967). See also Howard H. Schless, *Chaucer and Dante: A Revaluation* (Norman, OK: Pilgrim, 1984), pp. 29–30.

works, it is well worth making reference to them, since they indicate what was known and available in more general sources after 1250', highlighting in addition the degree to which 'such secondary lines of influence [as Scriptural commentaries and sermons] brought optics to a non-specialist audience' (pp. 30–33). By the time that Lydgate was writing in the fifteenth century, this non-specialist audience would have reached saturation point in terms of the dissemination of optics as a discursive formation. As such Lydgate's knowledge of optics was not specific to him, but diffused throughout the late medieval poetic tradition and the wider pan-European intellectual apparatus upon which that tradition drew, and throughout theological practice in the schools and in the pulpit. The specialist knowledge and the specialist discourse which both enabled and validated it thus passed into more common usage, to the point that even a poet such as Lydgate—for so long an emblem of fifteenth-century dullness—could be illuminated by it, and reflect it on to others.[31]

This reflection is enacted within the Temple itself by the knight's echoing of the narrator's language in his forlorn account of the lady's splendour and its effects upon him:

> But nov of nwe within his fire cheyne
> I am enbraced, so that I mai not striue
> To loue and serue, whiles that I am on lyue,
> The goodli fressh, in the tempil yonder
> I saugh right nov, that I hade wonder,
> Hou euer god, forto reken all,
> Myght make a thing so celestial,
> So avngellike on erthe to appere.
> For with the stremes of hir eyen clere
> I am Iwoundid euen to the hert,
> That fro the deth, I trow, I mai not stert.
> [...]
> Alas! when shal this tempest ouerdrawe,
> To clere the skies of myn aduersite,
> The lode ster when I [ne] may not se,

---

[31] Emily McCarthy's fascinating paper on light in the poetry of William Dunbar, delivered at this conference, perhaps reinforces Lydgatean reflection, or rather Lydgate's refraction of Chaucer's legacy into a focus upon *rhetorick swete*, as part of what Paul Strohm has termed the narrowing of the Chaucer tradition. See 'Chaucer's Fifteenth Century Audience and the Narrowing of the "Chaucer Tradition"', *Studies in the Age of Chaucer*, 4 (1982), 3–32. However, this narrowing is not necessarily due to an inability to grasp Chaucerian syncretism, but is due to an aesthetic position.

> It is so hid with cloudes that ben blake.
> [...]
> For in myn hert enprentid is so sore
> Hir shap, hir fourme, and al hir semelines,
> Hir port, hir chere, hir goodnes more & more
> [...]
> Mirrour of wit, ground of gouernaunce,
> A world of beaute compassid in hir face,
> Whose persant loke doth thurugh myn hert[e] race (ll. 574–756)

Aside from the recalls of the cloudy night and the lady's luminescence, the knight's complaint of how 'with the stremes of hir eyen clere / I am Iwoundid euen to the hert', reveals Lydgate's adhesion to the traditional account of love which passes from the eyes of the beloved into the eyes of the lover, and from thence to the fleshly tables of the heart.[32] Whether this process is intromissive or extramissive—or perhaps a fusion of the two, depending upon objective and subjective perspectives—is, however, beyond my present remit. Yet it is worthwhile comparing the account of the lady's beauty with the description of the Goddess *Natura* that Lydgate provides in *Reson and Sensuallyte*, an allegorical poem which many commentators believe was written following the *Temple of Glas*:

> Whan I beheld hir woman-hede
> And the beaute of hir face,
> The whiche abouten al the place
> Caste so mervelous a lyght,
> So clere, so percynge and so bryght,
> That the goddesse Proserpyne
> With al hir bryghte stonys fyne
> And hir ryche perles clere
> To hir beaute ne myght appere.
> They were so percyng and so chene,
> That I ne myghte nat sustene
> In hir presence to abyde,
> But went bak and stood asyde,
> Til at the last[e], in certeyn,

---

[32] See for example Andreas Capellanus, *De Amore* (*Andreas Capellanus on Love*), ed. and trans. P. G. Walsh (London: Duckworth, 1982), pp. 32–33, or the pool of Narcissus in the *Roman de la Rose*. A useful discussion is to be found in the opening chapter of A. C. Spearing, *The Medieval Poet as Voyeur* (Cambridge: Cambridge University Press, 1993), pp. 1–25.

I Forced me [onward] ageyn,
Hert and body, in sothnesse (*RS*, 212–27)[33]

If the mirroring structure of the poem finds the knight giving voice to the implicit complaint of the dreamer—as Davidoff posits—then it is balanced by the linguistic parallels between the description of the lady and that of the goddess. The lady herself describes Venus as 'persant & ful of light, / Of bemys gladsome, devoider of derknes, / Cheif recounford after the blak nyght, / To voide woful oute of her heuynes' (ll. 328–31), which echoes her own luminous description by the dreamer. And not only this but such language has an obvious message for the dreamer himself: by following the doctrine of *patientia* outlined by the Goddess he will become devoid of darkness, as Venus herself informs the lady (ll. 377–453). The knight likewise equates the lady and Venus in that 'the stremes of hir [the beloved's] eiyen clere' are echoed by 'the stremes of thi plesaunt hete [...] Whos bright bemes ben wasshen and of[t] wete / In the riuer of Elicon the well' (ll. 701–06).[34] Furthermore, the account of how Venus 'ye me hurten with your dredful myght / Bi influence of your bemys clere' (ll. 717–18) corresponds with the account of the pain inflicted by the streams of luminous beauty emitted by the lady. Ultimately the link between Venus and the lady is clarified by the goddess's revelation that 'I haue gyue hir of beaute excellence, / Aboue al othir in vertue forto shine' (ll. 1189–90), and her proclamation to the Knight that he should

> Thenk hou she is this wor[l]dis sonne & light,
> The sterre of beaute, flour eke of fairnes—
> Bothe crop and rote—and eke the rubie bright
> Hertes to glade Itroubled with derknes. (ll. 1208–11)

Lydgate thus appropriates the stock language of courtly love with which he was so familiar through his reading of Chaucer and the French *dits*

---

[33] Ernst Sieper, ed., *Lydgate's Reson and Sensuallyte*, EETS, e.s. 84, 89 (London: Kegan Paul, Trench, Trübner, 1901–03; repr. 1965), I, 6–7. The supposed proximity of the two poems inevitably revives the question of the *Temple*'s dating.

[34] The conclusion to be drawn from this image is that poetry provides a refracted form of pure love, in that Venus's light enters into and informs the transparent medium of Helicon's waters.

amoreux, and employs it in such a way that it upholds his poem's central focus upon light and darkness.[35]

Indeed, Chaucer casts a long shadow over Lydgate's poetry and commentary upon it, and the present study is no exception.[36] However, rather than retreat to the once traditional critical portrait of Lydgate as a slavish, second-rate Chaucerian, it is worth considering briefly how Lydgate incorporates Chaucerian memories and echoes within his poem, the purpose that they serve, and how he develops them within the remit of his own poetics. Recent Lydgate studies – often drawing on the restitutionism of David Lawton and James Simpson – have sought to explain Lydgate's Chauceriana by pointing to the accretive nature of late medieval textuality, whereby the post-Romantic concept of authorial originality proves anachronistic, whilst at the same time pointing (paradoxically) towards Lydgate's own originality, particularly in relation to his longer works, such as the *Fall of Princes*, the *Troy Book* and the *Siege of Thebes*.[37] Lydgate's courtly lyrics, on the other hand, have been relatively ignored. Perhaps this is because they are less amenable to new historicist excavation, which privileges the political and the ideological, or perhaps it is because their Chaucerian inheritance is more in evidence.[38]

---

[35] On Lydgate's reading of French poetry see Susan Bianco, 'A Black Monk in the Rose Garden: Lydgate and the *Dit Amoureux* Tradition', *Chaucer Review*, 34 (1999), 60–68.

[36] We might trace this shadow back to Joseph Ritson's influential charge that Lydgate 'disgraces the name and patronage of his master Chaucer'. See *Bibliographia Poetica: A catalogue of English poets of the twelfth, thirteenth, fourteenth, fifteenth, and sixteenth centurys, with a short account of their works* (London: Roworth, 1802), p. 88. However, Lydgate's perceived comparative inferiority was noted as far back as Puttenham and Sidney, and in many ways is dependent upon Lydgate's own veneration of Chaucer. On Lydgate's critical heritage see Renoir, pp. 1–31, Pearsall, *John Lydgate*, pp. 1–21, and Nigel Mortimer, *John Lydgate's Fall of Princes: Narrative Tragedy in its Literary and Political Contexts* (Oxford: Clarendon Press, 2005), pp. 1–24. See also my '"Disgraces the name and patronage of his master Chaucer": Echoes and Reflections in Lydgate's Courtly Poetry', in *Standing in the Shadow of the Master: Chaucerian Influences and Interpretations*, ed. Kathleen A. Bishop (Newcastle: Cambridge Scholars Press, 2010), pp. 2–27, which develops certain of the points raised by my earlier study, 'The Marginalization of John Lydgate', *Marginalia*, 1 (2005) <http://marginalia.co.uk/journal/05margins/rossiter.php> [accessed 3 March 2013].

[37] See Lawton's seminal 'Dullness and the Fifteenth Century', *ELH*, 54 (1987), 761–99 and Simpson's *Reform and Cultural Revolution*, pp. 34–67.

[38] However, Mitchell's study argues for engagement with contemporary politics at court as being at the heart of the *Temple*.

Chaucer was, after all, praised and revered as a 'noble Rhetor' in the fifteenth century, not least of all by Lydgate himself in poems such as *The Floure of Curtesye* and the *Life of Our Lady*.[39] Whatever the reason, it seems somewhat uneven to overlook those courtly poems which so often reveal what is best in Lydgate; that is, his collation and refiguring of preceding materials within different contexts and semantic codes.

The temple itself is a piece of Chaucerian architecture, appearing as it does in the dreamscape of the *House of Fame*:

> But as I slepte, me mette that I was
> Withyn a temple ymad of glas,
> In which ther were moo ymages
> [...]
> For certeynly, I nyste never
> Wher that I was, but wel wyste I
> Hyt was of Venus redely (ll. 119–30)

However, Lydgate employs the temple differently from Chaucer, as we have seen; its material serving to blind the dreamer following the darkness of the dream-frame, which in turn serves to acclimatize his vision to the temple's interior decoration and the dazzling beauty of the lady and the goddess. Similarly, Lydgate echoes the forlorn narrator of the Chaucerian dream-vision. Yet whereas the narrator of the *Book of the Duchess* complains that he cannot sleep due to his melancholy ('I have gret wonder, be this lyght, / How that I lyve, for day ne nyght / I may nat slepe', 1–3), Lydgate's narrator takes to his bed because of it in the corresponding lines of his poem: 'For thought, constreint, and greuous heuines, / For pensifhede, and for heigh distress, / To bed I went nov this othir nyght'. Likewise, at the close of the poem, Lydgate's narrator declares that 'I purpose here to maken & to write / A litil tretise, and a processe make / In pris of wommen' (ll. 1379–81), as Chaucer declares he will at the close of *Troilus and Criseyde* (V. 1777–78), and as he is enjoined to do by Alcestis in the Prologue to the *Legend of Good Women* (F 435–41 / G 425–31). Again the contexts differ: Chaucer proposes his treatise at the end of *Troilus and Criseyde* on account of his representation

---

[39] See Julia Boffey, 'The Reputation of Chaucer's Lyrics in the Fifteenth Century', *ChauR*, 28 (1993), 23–40, and David R. Carlson, 'The Chronology of Lydgate's Chaucer References', *Chaucer Review*, 38 (2004), 246–54. Lydgate refers to Chaucer as 'noble Rhetor' in the *Life of Our Lady* (II. 1629). See Joseph A. Lauritis et al., eds, *A Critical Edition of John Lydgate's Life of Our Lady* (Pittsburgh, PA, 1961).

of Criseyde's unfaithfulness, whilst 'Geffrey' in the Prologue to the *Legend of Good Women* is commanded to amend the wrongs he has done women through his poetry, as a forfeit, rather than risking the vengeance of Cupid. Lydgate's narrator, on the other hand, proposes his treatise '[b]icause I had neuer in my life aforne / Sei[n] none so faire, fro time that I was borne' (ll. 1376–77); his reasons for writing are 'oonli for her sake' (l. 1381).

The *Temple of Glas*, which 'was of Venus redely', also draws on the temple of Venus in the *Parliament of Fowls*. However, whereas Lydgate's narrator begins in darkness and awakens into the dazzling dream of the temple, Chaucer's narrator moves from light to darkness in the frame, and from darkness to light in the dream:

> The day gan faylen, and the derke nyght,
> That reveth bestes from here besynesse,
> Berafte me my bok for lak of light
> […]
> I saw a temple of bras ifounded stronge
> […]
> Derk was that place, but afterward lightnesse
> I saw a lyte, unnethe it myghte be lesse
> (*PF*, 85–87, 231, 263–64)

Chaucer's dream is born of reading; indeed he repeatedly advertises the intertextual nature of his dream poetry. In this instance the narrator has been reading 'Tullyus of the Drem of Scipioun' (*PF*, 31), that is Cicero's *Somnium Scipionis*, with Macrobius' commentary. In the *Book of the Duchess* the narrator reads the tale of Ceyx and Alcione from Ovid's *Metamorphoses* (*BD*, 44–369), refracted through Machaut's *Dit de la fonteinne amoreuse* (ll. 543–698).[40] Lydgate's poem is clearly also dependent upon his reading, in particular his reading of Chaucer, albeit his narrator takes no book to his bed. In fact, Chaucer's assertion that 'out of old bokes, in good feyth, / Cometh al this newe science that men lere' (*PF*, 24–25) may be seen to illustrate the accretive construction of Lydgate's temple, drawing as it does upon both 'old bokes' and 'newe science'. And whilst there is neither the time nor the space here to trace every single Chaucerian inflection which appears in Lydgate's poem, we can nevertheless see from those instances which we have touched upon that Lydgate is following the late medieval tradition of material

---

[40] See James I. Wimsatt, *Chaucer and His Contemporaries: Natural Music in the Fourteenth Century* (Toronto: University of Toronto Press, 1991), pp. 83–84.

appropriation and reallocation—what we might think of as poetic recycling.

However, Lydgate makes a clear departure from the Chaucerian model at the close of his poem in terms of the break between the dream proper and the frame, a break which is effected by the same *chiaroscuro* that characterized the initial movement into the vision. The dream concludes with a hymn of praise to the goddess, sung by all those present in the temple, a hymn which is suffused with luminescent language:

> 'Fairest of sterres, that, with youre persant light
> And with the cherisshing of youre stremes clere,
> Causen in loue hertes to ben light,
> Oonli thurugh shynyng of youre glade spere:
> Nou laude and pris, o Venus, ladi dere,
> Be to your name, that haue withoute synne
> This man fortuned his ladi forto wynne.
>
> [...]
>
> O myghti goddes, daister after nyght,
> Glading the morov whan ye done appere,
> To voide derknes thurugh fresshnes of your sight,
> Oonli with twinkeling of youre plesaunt chere:
> To yov we thank, louers that ben here,
> That ye this man—and neuer forto twyn—
> Fortuned haue his ladi forto wynne.'
>
> And with the noise and heuenli melodie
> Which that thei made in her armonye
> Thurugh oute the temple, for this manes sake,
> Oute of my slepe anone I did awake,
> And for astonied knwe as tho no rede;
> For sodein chaunge oppressid so with drede
> Me thought I was cast as in a traunce:
> So clene away was tho my remembraunce
> Of al my dreme, wher-of gret thought & wo
> I hade in hert, & nyst what was to do,
> For heuynes that I hade lost the sight
> Of hir that I, all the long[e] nyght,
> Had dremed of in myn auisioun:
> Whereof I made gret lamentacioun (ll. 1341–75)

The melodious sound awakens the dreamer to the darkness of melancholy, as he laments the loss of the light which had so ravished his sense, which again may be said to exemplify the instantaneous switch between actuality and potential in Aristotelian optics. Lydgate seems to willingly frustrate the expectations of his audience by refusing to endorse what Davidoff terms 'the movement from need to fulfilment' (p. 122), which is characteristic of the late medieval dream-vision. In doing so he may be seen to expand the audience's *Erwartungshorizont* ('horizon of expectations').[41] Davidoff concludes that the poem concerns 'the light-giving property of words [...] a statement of how writing poetry can dispel a dreamer's darkness', although she also concedes that 'it operates on several levels' (pp. 124–25). I would certainly endorse this latter reading, not as a means of avoiding any definite judgment of the poem's meaning, but in order to celebrate its multiplicity or *polysemia*. Lydgate was writing during a period in which light played multiple roles throughout various disciplines, and as such it would be erroneous to suggest that illumination is confined to a single metaphorical vehicle, especially given the poet's aforementioned inclination towards encyclopaedic accretion. Indeed, it is surely necessary when reading a poem such as the *Temple of Glas* to dispel the negative undertone which usually accompanies the term encyclopaedism. That is, we might think of it not so much as an indiscriminate accretion of *materiae*, as a deliberate means of generating an interpretative multiplicity.

This multiplicity, moreover, does not mean that the discursive variations upon the theme of light have no correlation. The language of optics, theology and poetry inform one another within the poem. The opening spiritual and physical darkness of the dreamer-to-be is illuminated by an act of blinding that is actually a spiritual and physical reflection, which gives way to the vision of the lady who is 'this wor[l]dis sonne & light'. This description transforms the lady, like Chaucer's Griselda, into a *figura Christi*: 'Ego sum lux mundi: qui sequitur me, non ambulat in tenebris, sed habebit lumen vitae' ('I am the light of the world: he that followeth me shall not walk in darkness, but shall have the light of life', John 8. 12).[42] The physically blinding light, redolent of the Damascene conversion, thus prepares for what is a beatific vision—a glimpse of the *lux mundi*—which restores the equally Pauline sense of

---

[41] The term is central to the reception theory of H. R. Jauss. See *Toward an Aesthetic of Reception*, trans. Timothy Bahti (Brighton: Harvester, 1982), p. 44.

[42] On Griselda as *figura Christi* see Marga Cottino-Jones, 'Fabula vs. Figura: Another Interpretation of the Griselda Story', *Italica*, 50 (1973), 38–52 (p. 41). On the importance of Griselda as a referent in the *Temple of Glas* see Scanlon, 'Lydgate's Poetics', pp. 80–91. On the tale of Griselda as it passed from Italian to Latin to English, see chapters 4 and 5 of my *Chaucer and Petrarch*.

being '[r]auysshid in spirit'; albeit a vision which 'conflates the erotic and the religious'.[43] This conflation, characteristic of *amour courtois*, is predicated upon the luminescent vision of the *donna angelicata*, and raises poetic illumination to a status equal to that of optical and theological equivalents, as Venus's 'bright bemes ben wasshen and of[t] wete / In the riuer of Elicon the well'. Each discourse complements the other, and whilst the poem appears to signify light as a means of unifying science, theology and poetics, none of these is allowed to dominate. I find that Lindberg's explanation of Roger Bacon's synthesis provides a useful analogue for thinking about Lydgate's movement towards what might be termed a pure allegory predicated upon a pure light or *lux*, of which we receive only the faintest *lumen*. He argues that

> If the Baconian synthesis lacked depth and reflected an uncritical belief in the unity of knowledge, that by no means interfered with its wide acceptance in the thirteenth century and its continuing influence during the later Middle Ages. (p. 116)

We might say the same for Lydgate's poem, which maintained its popularity throughout the fifteenth century and into the sixteenth: it is predicated upon 'belief in the unity of knowledge'. And if we cannot fully decipher its 'plein [...] significaunce' this is perhaps because now, from an allegorical perspective, we see through Lydgate's glass darkly, whereas audiences then had more chance of seeing face to face.

---

[43] Scanlon, 'Lydgate's Poetics', p. 89.

Divine Light and Spiritual Intoxication:
Symeon the New Theologian's Image of Penitence as a Mystical Winepress

Hannah Hunt

Among Eastern Christian writers, St Symeon the New Theologian provides the most significant analysis of light as an indication of spiritual charism. This paper focuses on one detailed image as an example of his skill in conveying spiritual teaching through literary devices which evoke visual effects. Light permeates his writing, but the image found in *Catechesis* 23 is especially striking.[1] Here, after an extended appraisal of spiritual self-examination, there is the extraordinary image of repentance being like sunlight passing through a glass of wine, set in the framework of the mystical winepress and the relationship between divine light and sober inebriation. This may seem to be a small point but it raises a number of fascinating issues: is it possible—or even desirable—to trace a coherent theological use of concepts of luminosity within Symeon's writings in general? Is there one plausible exegesis of the wine, the sunlight and the drinker in this particular image? Symeon's use of light here is both within an established biblical, patristic and philosophical tradition but also articulates his peculiarly individual, at times almost playful, mode of expression. Bearing in mind the fact that the majority of extant texts are sermons written for what appears to be a fairly wilful collection of monks, it seems clear that some aspects of his style of writing may derive from the need to keep his congregation attentive. He manages to synthesize sophisticated theological insights with great literary beauty. His passion for light both articulates standard biblical light/dark imagery and also foreshadows the fourteenth century hesychasts' focus on the uncreated light (Luke 9. 28–36) as revealing the 'energies' of God.[2] Both of these intellectual and spiritual frameworks need to be explored in order

---

[1] Symeon the New Theologian, *The Discourses*, trans. C. J. de Catanzaro (New York: Paulist Press, 1980), pp. 254–260. All further references will be to this translation, abbreviated as *Discourses*.

[2] For full details of the hesychasts and their method of prayer, see I. Hausherr, 'La Méthode d'oraison hésychaste', *Orientalia Christiana Analecta*, 9 (1927), 100–210.

to help place Symeon in the context of eleventh century monasticism in Constantinople.

The Bible's message begins and ends with light or its absence; Genesis 1. 1–2 describes the creation of the world as a movement from 'darkness over the face of the void' to the separation of day and night in the command: 'Let there be light' (so effectively expressed in musical terms in the opening of Haydn's 'Creation'). For the Hebrew people, God's mastery over the universe was articulated by the movement from chaos to order, and light breaking through darkness symbolises how this happened. Chapter 22 of Revelation describes the perpetual conquering of night, as God's own light will replace the need for any 'lamp or sun', with Jesus as the 'bright star of dawn' (Revelation 22. 5, 16). Both the Old and New Testaments use imagery of light to indicate the presence of God and salvation, sometimes as a simple polarity between light as good and life-bearing and darkness as death-dealing and corrupt; at a metaphorical level, spiritual sight and blindness bring this dichotomy into human expression. The apostle Paul, who himself experienced conversion to Christianity through a period of apparent physical blindness (Acts 9. 3–9), writes frequently of light as in opposition to darkness, with no ambiguity about which is the one to choose, urging:

> Do not unite yourself with unbelievers; they are no fit mates for you. What has righteousness to do with darkness? Can light consort with darkness? (II Corinthians 6. 14–15).

The correlation between Christian virtues and light are found also in Ephesians 5. 7–9:

> For though you were once all darkness, now as Christians you are light. Live like men who are at home in daylight, for where light is, there all goodness springs up, all justice and truth.

In the Johannine literature, light/darkness dualism is inferred from the opening lines of the fourth gospel, which hints at the creation story depicted in Genesis, with 'the light of men' (John 1. 4) shining in a darkness which cannot overcome it, and John the Baptist, known in the Eastern Christian world as John the Forerunner, being not the light itself but the witness to it in the person of Christ (John 1. 8).

In addition to his own creative wordplay, Symeon's texts make frequent reference to images in both testaments of light and darkness, sometimes by direct citation and elsewhere by allusion.³ It was axiomatic of patristic writing that adherence to a tradition is seen as having more value than innovation, and this attitude is still prevalent at the time of Symeon. Whilst remaining faithful to the commonly received interpretation of the significance of light, Symeon was also able to draw on a millennium of subtle modifications and different examples of theological writing on light. Constantinople at the time housed many great churches which used the fall of natural and artificial light on mosaics and icons, so there were physical as well as symbolic reminders of the importance of light for the Christian.⁴ Symeon was not concerned with contributing anything original to an established use of light as a general indicator of spiritual awareness and maturity. However, his interpretation of personal encounters with divine light as the basis for the whole mystical framework, as indeed the starting point of the spiritual quest, was radical and challenging and needs to be examined within the context of his *vita* in order to contextualise the image in *Catechesis* 23. Apart from anything else, he stressed an elitism: there is, he argued, only one in a thousand (or even ten thousand) who can progress from baptism to achieve deep mystical contemplation.⁵ Whether this is mere solipsism or evidence of a systematic reworking of patristic tradition is explored elsewhere.⁶

From the earliest Christian times writers drew on light as an image for Christ and for the right way to live. Hastings notes the presence of the phrase 'light from light' in the earliest attempts to put together a common creed at Nicaea in 325.⁷ The significance of this is the contemporary urgency to separate the developing Christian church from strongly supported Christological heresies such as Arianism. If in Biblical times, God's people needed to be encouraged to chose light over

---

³ See, for example, *Catechesis* 28 (4), *Discourses*, p. 298, which refers to John's gospel and also Luke 1. 79; section 10 of the *Catechesis* refers to 8. 12 and 10. 7, 9, *Discourses*, p. 303; references to Galatians 3. 27 and Ephesians 5. 8 in *Catechesis* 19, *Discourses*, p. 229; see also *Catechesis* 34 (12), *Discourses*, p. 357.

⁴ *Oxford Dictionary of Byzantium*, ed. A. P. Kazhdan, 3 vols (Oxford: Oxford University Press, 1991), II, 1227–28.

⁵ *Hymn* 50, 152–254, in St Symeon the New Theologian, *Hymns of Divine Love*, intro. and trans. G. A. Maloney (Denville, NJ: Dimension Books, 1975), p. 253.

⁶ H. Alfayev, *St Symeon the New Theologian and Orthodox Tradition* (Oxford: Oxford University Press, 2000).

⁷ Entry on 'light and darkness' in *The Oxford Companion to Christian Thought*, ed. A. Hastings, A. Mason, and H. Pyper (Oxford: Oxford University Press, 2000), p. 390.

darkness, likewise the fourth and fifth centuries of the Christian era found the recorders of doctrine reliant on such motivation in order to refute heresy.

To be a Christian was to walk in the light, and this is readily seen in the use of light in the liturgies of Christian initiation. Baptism swiftly became crucial to Christian discipleship and by the time of Symeon the process involved light, water and penitential tears.[8] From the time of Justin Martyr and Ignatius in the late second century, illumination was the chosen metaphor for the process of initiation into the Christian church through baptism, reflecting initiation ceremonies from other, pre-Christian, religious traditions.[9]

So light appears in the Bible and the early church as an image of God separating his chosen people from a place of darkness representing evil and error. There are also specific points in the life of Christ where Jesus himself is associated with light, and one of these in particular is developed into a liturgical feast as well as the source of much influential art work. This ratifying by the church of a light-filled encounter gives great weight to the role of light in Christian experience. The sixth of August became designated the Feast of the Transfiguration (one of twelve major feasts in the Orthodox Church) by the start of the eighth century.[10] Luke 9. 29 (cf. Matthew 17. 1–8) describes Jesus' transformation by an 'uncreated light', on Mount Tabor, when his face and clothes became dazzlingly luminous as he talked with Moses and Elijah, and, like them, received a commission from God. This encounter with God on Mount Sinai, from which less 'illuminated' mortals such as his disciples must be protected, inspired a major mosaic in St Catherine's monastery in Sinai.[11] The liturgical texts for this feast return constantly to the theme of light, inviting those present to 'mystically behold Christ shine as lightning with the rays of divine splendour' and an invocation to the 'blessed Master [...] Thou dwellest in unapproachable light'.[12] The transfiguration is the link to the

---

[8] See H. Hunt, *Joy-Bearing Grief: Tears of Contrition in the Writings of the Early Syrian and Byzantine Fathers* (Leiden: Brill, 2004), p. 203.

[9] *Oxford Dictionary of Byzantium*, II, 1227.

[10] *Oxford Dictionary of Byzantium*, III, 2104.

[11] *Oxford Dictionary of Byzantium*, III, 2104 and see also W. Treadgold, *A History of the Byzantine State and Society* (Stanford: Stanford University Press, 1997), p. 273 and pp. 835–36 for the endurance of this image into the thirteenth and fourteenth centuries.

[12] *Festal Menaion*, trans. Mother Mary and K. Ware (South Canaan, PA: St Tikhon's Seminary Press, 1990), pp. 468 and 492.

hesychasts of the fourteenth century, for whom Symeon is seen as a key prompt. Gregory Palamas, Archbishop of Thessalonike between 1347 and 1359, built on the Athonite practice of contemplative prayer to construct a philosophical argument which proved (like much of Symeon's teaching) to be politically controversial. He sought to distinguish between the unknowable essence of God and the accessible 'energies' of God which were manifest by the 'uncreated light' of the Transfiguration. There are some parallels but also significant differences from Symeon's reading of light as a means of understanding God. The significance of the transfiguration for Symeon was that it constituted *theosis* for the spiritual traveller rather than articulating Christological doctrine.[13] Also his focus is primarily on the experiential nature of light, however it is defined.[14]

Whereas Palamas sought to distinguish at a philosophical level between the essence and energies of God (the essence being unknowable; the energies in the form of deifying light being accessible through contemplative prayer), Symeon used light to indicate the presence of God in a human person. For Palamas, light was the means by which a contemplative experienced God. He taught that through the right type of hesychastic prayer one could experience God in the uncreated light of the transfiguration of Jesus, the so-called Taboric light. Although Symeon is frequently linked with hesychasm, for him light was not so much the means by which an encounter with God took place as the signifier of God's grace in an individual. This is placed within the context of a golden chain of spiritual charism, which is passed from spiritual father to spiritual son. Symeon's use of light is extremely pervasive and varied, hence the decision to focus in detail on one specific example on this occasion. In the autobiographical accounts of visions of light, Symeon insisted on light being a demonstration of holiness or spiritual illumination especially in one who is destined to lead others to light; elsewhere he used light to indicate God's presence, or in a number of metaphorical senses some of which draw on biblical imagery of light. For this eleventh-century abbot of St Mamas in Constantinople, light not only shaped his own spiritual development but made sense of the immediacy of a monk's encounter with the living God. This emphasis on the immediate and experiential was at the heart of his theology. Symeon thus developed the existing understanding of the transformative quality of

---

[13] See comment by A. Cameron, *The Byzantines* (Oxford: Blackwells, 2006), pp. 110–111 on how this related to hesychasm.

[14] Alfayev, *St Symeon the New Theologian*, p. 226.

Taboric light to one in which 'the whole of human nature, including the intellect, the soul, and even the body, is transfigured by the divine light'.[15]

Light, whether as a source of physical or spiritual illumination, as reflected, dispersed or metaphorical, takes many guises in Symeon's writings, and before looking at the particular image of sunlight as repentance, it is worth looking briefly at the scope and range of light within Symeon's works as a whole, in order to suggest the particular place of the light of repentance. At its most domestic is the image of firelight; more commonly he develops the image of the soul as a lamp which incorporates a divine fire, as in: 'When the lamp of the soul, that is the mind, has been kindled, then it knows that a divine fire has taken hold of it and inflames it.'[16] The association of the soul with light may owe something to the Macarian tradition: in *Homily* 1, 2. 2 Macarius talks of the soul as 'completely illumined with the unspeakable beauty of the glory of the light of the face of Christ' through which it is made 'a participator of the Holy Spirit'.[17] The consciousness of the soul's illumination by God is typical of Symeon. But many of his images are more cosmic: note the image of the sunbeam penetrating the darkened house in *Catechesis* 15.[18] The final verses of Revelation are evoked in *Catechesis* 22 where he suggests that 'when the visible sun sets, this sweet light of the spiritual star takes its place'.[19] Chapter 3. 54 contains the image of a man inside a house opening a window and seeing a flash of

---

[15] Alfeyev, *St Symeon the New Theologian*, p. 237.

[16] *Catechesis* 15 (3), *Discourses*, p. 195; cf *Catechesis* 33 (1), *Discourses*, p. 339; *Catechesis* 4(16), *Discourses*, p. 88; *Catechesis* 33 (1), *Discourses*, p. 339; and *Catechesis* 33 (2), *Discourses*, p. 341. Alfeyev, *St Symeon the New Theologian*, p. 239 notes how close Symeon is in his handling of images of fire to the author of the Makarian Homilies, which talk of the 'kindling of the soul by the fire of God'.

[17] Pseudo-Macarius, *The Fifty Spiritual Homilies and the Great Letter*, trans. G. Maloney (New York: Paulist Press, 1992), p. 37. In this Homily, Macarius also talks of the 'spiritual eyes of light', which recall some of Ephrem the Syrian's language of the luminous eye of the soul and of faith, eloquently explained in S. Brock, *The Luminous Eye* (Kalamazoo, MI: Cistercian Publications, 1985).

[18] *Discourses*, p. 194.

[19] *Discourses*, p. 248.

lightning, which is compared to the soul enclosed in the senses, which cannot bear the radiance of this unbearable light.[20]

The light may be domestic and familiar, or cosmic and uncontrollable; references to it may form a passing aside, a substantial and developed metaphor, or a mixture of several images. For example, in *Hymn* 33, he combines an image of God as the source of illumination with the physical process of kindling light as a spiritual discipleship. Here he states the purpose of ascetic practices as being: 'only in order to partake of the divine light, light a lamp, so that we may bring out souls as a single candle to the inaccessible light.'[21] In common with the Johannine and Pauline passages referred to above, the light is usually described as being in contrast to the surrounding darkness, as in *Catechesis* 19 (4): 'If then the light of the commandments of Christ shines on [the soul of man], it finds itself in the infinite light of His gracious Godhead' which brings 'unutterable and unending joy'. By contrast, being 'shrouded in the darkness of sins' means the soul is 'in unending darkness mingled with fire'.[22]

But in addition to using the brightness and luminescence of light as a metaphor for spiritual growth and maturity, Symeon also describes its ability to reflect and disperse, which enables him to use light to explain doctrinal teachings such as the Incarnation and Trinity, where a sense of interconnectedness of apparently diverse qualities needs to be explained. He shares this tradition with the Cappadocian Fathers, and also Evagrios, who talks of the light of the Holy Trinity and the luminosity of the dispassionate person at prayer.[23] Evagrian light mysticism is also found in Syriac writers such as Isaac of Nineveh and John of Dalyatha and there is evidence in other aspects of Symeon's writing (such as his texts on *penthos*) that he had either read these writers or was drawing from a common pool of imagery and doctrine. In the case of Symeon, the associations are varied rather than consistent. Light may be connected to the

---

[20] Symeon the New Theologian, *The Practical and Theological Chapters and the Three Theological Discourses*, trans. P. (= J. A.) McGuckin (Kalamazoo, MI: Cistercian Publications, 1982), pp. 87–88.

[21] See the explanation given by H. Alfeyev, 'The Patristic Background of Symeon the New Theologian's Doctrine of Light', *Studia Patristica*, 32 (1997), 229–238 (p. 238).

[22] *Discourses*, p. 229. For many more instances of light imagery in his writings, see Alfeyev, St Symeon the New Theologian, p. 238 and the whole of chapter X. 3 (pp. 226–41).

[23] For a discussion of light in the earlier Eastern Christian tradition, see Alfeyev, *St Symeon the New Theologian*, esp. pp. 226 and 203.

person of Christ and frequently the Holy Spirit, but may also be used to describe the unity of the Trinity.[24] Predominantly, he followed patristic tradition by using metaphorical language to affirm the undivided nature of the Trinity, with the concept of light as a defence of traditional teaching: 'If you speak of 'light' (I John 5. 8), then both each Person is light and the Three are one light.'[25]

The *Third Theological Discourse* gives an extensive excursus into doctrinal discussion, employing light imagery. He argues for the unity of the Trinity and the 'permanent identity of the three hypostases within the external glory and the inseparable oneness'.[26] The text contained standard arguments and gospel citation and a typically personal invocation to the Holy Spirit as the 'remedy of sin and gateway of all repentance'.[27] He briefly elaborated on the common analogy of Trinity as the divergent waters of a river coming from a single stream, and towards the end of a rationally argued philosophical discourse, he introduces an extensive and ecstatic eulogy of God as light, in which the images come so thick and fast it is almost impossible to analyse them accurately. First he describes as light each of the persons of the Trinity, then suggests they are 'one single light as they are simple, non-composite, timeless, eternal, and possessed of the same honor and glory'. Then he refers to God as being the source of all light, with everything 'given to us as arising from the light'. Finally, he lists a series of concrete and abstract nouns – life, life, immortality, the bridal chamber, the garments of the saints, his resurrection – as being light, culminating in a catena of scriptural references describing 'the comforter, the pearl, the seed of mustard, the true vine' as light.[28] Although frequently his approach is apophatic in this homily, he uses the language of 'energies' (prefiguring Palamas), to develop his discussion about light from an abstract desideratum for Christian living to a description of divine qualities: 'His goodness is light; his compassion [...]

---

[24] 'The simple light is Christ. So he who has His light shining in his mind is said to have the mind of Christ.' *Catechesis* 33 (1), *Discourses*, p. 340; cf. *Catechesis* 2 (15): 'we may see the divine Light, Christ Himself, and possess Him abiding in us.' *Discourses*, pp. 58–59.

[25] *Catechesis* 33 (8), *Discourses*, p. 344.

[26] Symeon, *The Practical and Theological Chapters*, trans. P. (=J. A.) McGuckin, p. 135.

[27] Ibid., p. 136.

[28] Ibid., pp. 138–39.

his mercy, his embrace, his watchful care are light. His sceptre is light, his crook, and his consolation.'[29]

Aspects of light not only explain the unity and co-inherence of the Trinity, but also bridge humanity and divinity. Light is depicted as the means by which humanity can perceive and be touched by God, how 'those who are dark become light in wondrous fashion as they draw near to the Great Light'.[30] Drawing near to the source of light in this way resembles the mediation of the incarnation, with light representing the divine spark shared by Creator and created.[31] This insistence that light is the point of contact between God and human creation explains why for Symeon the conscious awareness of God was so crucial in the life of a Christian 'athlete', and why the polyvalent meanings of light work so well to explain this encounter.

Before approaching the handling of light in *Catechesis* 23, it is worth exploring the significance of light in Symeon's own life, as expounded in some of his autobiographical writings as well as the *Vita* composed by Nicetas Stethatos.[32] We also need to be aware that he draws on, even some would argue, synthesizes, two significant strands of theological thought, the light-filled Philokalic spirituality from Origen onwards and the affective heart-centered spirituality of such Syrian writers as Ephrem, Isaac, John Hazzaya and John of Dalyatha, as mentioned above.[33] But whilst Symeon is consciously part of a continuum of mystical thought, his *vita* constantly demonstrates that it is the intuitive and experiential on which he relied for confirmation of his beliefs about human engagement with God.

Scholars argue about the precise dating of Symeon's life.[34] Most place him as living between 949 and 1022, with the first vision of light taking place when he was in his early twenties and still living and working in the imperial court. This experience is described in detail in *Catechesis* 22,

---

[29] Ibid., p. 139.
[30] *Catechesis* 17 (4), *Discourses*, p. 207.
[31] *Catechesis* 18 (10), *Discourses*, p. 216.
[32] *Vie de Syméon le Nouveau Théologien*, ed. I. Hausherr, trans. G. Horn, *Orientalia Christiana*, vol. 12, no. 45 (1928).
[33] This area is explored in some depth in J. A. McGuckin, 'Symeon the New Theologian's Hymns of Divine Eros: A Neglected Masterpiece of the Christian Mystical Tradition', *Spiritus: A Journal of Christian Spirituality*, 5.2 (2005), 182–202 (pp. 187ff.).
[34] See H. Hunt, *Joy-Bearing Grief*, pp. 174–75 for a discussion of variants.

lines 88-104, but references to visions in the plural appear in the *Hymns of Divine Love*, especially number 25, and in more general terms in the *Fifth Ethical Discourse*. This first account employs a conventional affectation of referring to himself in the third person as 'George', whom he portrays as a handsome and worldly man in need of guidance and a way forward:

> One day, as he stood and recited, 'God, have mercy upon me, a sinner' (Luke 18. 13), uttering it with his mind rather than his mouth, suddenly a flood of divine radiance appeared from above and filled the room.

The rest of the event is presented with further biblical echoes, and is accompanied by tears expressive of both joy and repentance. He continues:

> In a wonderful manner there appeared to him, standing close to the light, the saint of whom we have spoken, the old man equal to the angels who had given him the commandment and the book.[35]

The 'old man' in question is his spiritual father, the Studite monk Symeon, usually referred to as Eulabes to distinguish him from his more famous spiritual son. In this account, Symeon is quite specific and original; other writers describing visions of light often associate them with angels or possibly the Mother of God.[36] Also notable are that from the outset, the vision is inspired by scripture; also that he is uttering the words of repentance 'with his mind' (in other words, this is an inner, metaphysical encounter, similar to that of the hesychasts, however graphically it is described and experienced.) Yet it is something which engrosses his whole person: there is a human integrity in the spiritual experience. Alfeyev sees a parallel to Diadochos' sense of the intellect becoming translucent when energised by divine light, and notes: 'The experience of Symeon proves that the whole of human nature, including the intellect, the soul and even the body, is transfigured by the divine light.'[37]

Above all, the overt connection between this first encounter and his spiritual father, Symeon Eulabes, is crucial in linking Symeon's doctrine of illuminated repentance to a man whose power to remit sin comes from personal charism and not ordination. This elevation of the power of spiritual illumination underpinned Symeon's understanding of *theosis* and

---

[35] *Catechesis 22, Discourses*, pp. 243–44.
[36] This is discussed in more detail in Alfeyev, *St Symeon the New Theologian*, p. 232.
[37] Alfeyev, 'The Patristic Background', pp. 232 and 238.

redemption. The significance of 'George' being still in the world emphasises that illumination must come first, as the catalyst prompting a departure from the world, rather than constituting a reward for years of spiritual discipline. Krivocheine explains that the vision of light is both the end of the ascent and part of the progression.[38] McGuckin eloquently suggests that illumination is given at the outset of the spiritual journey, in order to 'call chosen disciples [...] out of darkness and into the fire of love', and goes on to argue that *theosis* is 'not a Next Age phenomenon but the essence of the spiritual life on earth as it ought to be lived'.[39]

Connected to this is the imperative stressed continually by Symeon that you cannot speak with any authority about something you have never experienced. This is stridently expressed in the *First Theological Chapter*:

> If a man cannot feel intuitively that he has put on the image of our heavenly Lord Jesus Christ, man and God, over his rational and intellectual nature, then he remains but flesh and blood. He cannot gain the experience of spiritual glory by means of his reason, just as men who are blind from birth cannot know sunlight be reason alone.[40]

This uncompromising assertion is echoed in the *Fifth Ethical Discourse*, where he suggests that in putting on Christ, the baptised had put on God himself, and, in the context of a furious debate with Stephen of Nicomedia,[41] he warns against those who are not aware of this:

> He then who has put on God, will he not recognize with his intellect and see what he has clothed himself with [...] Only the dead feel nothing when they are clothed, and I am very much afraid that those who say such things are the ones who are really and truly dead and naked.[42]

---

[38] B. Krivocheine, *In the Light of Christ* (Crestwood, NY: St Vladimir's Seminary Press, 1986), pp. 225–26.

[39] McGuckin, 'Symeon the New Theologian's Hymns of Divine Eros', p. 192.

[40] Section 1, 53, in Symeon, *The Practical and Theological Chapters*, trans. P. (=J.A.) McGuckin, p. 46.

[41] See Symeon, *The Practical and Theological Chapters*, trans. P. (=J.A.) McGuckin, pp. 17–18.

[42] St Symeon the New Theologian, *On The Mystical Life*, vol. II (*On Virtue and Christian Life*), trans. A. Golitzin (Crestwood, NY: St Vladimir's Seminary Press, 1996), p. 46. Cf. from the same text: 'No one can say anything unless he has first seen the light with the eyes of his soul and knows precisely its illuminations and activities as they occur within himself', pp. 52–3, and *Hymn* 21, line 160, where Symeon suggests that if you have not felt the light you should not dare to speak of

Returning briefly to Symeon's own *vita*, he was established as a novice, now firmly under the guidance of Symeon the Studite, (whose own monastic tradition features much talk of light and vision)[43] and again experienced visions of light in the context of begging for mercy for his sins. In *Catechesis* 16, he writes more honestly in the first person:

> I fell prostrate on the ground, and at once I saw, and behold, a great light was immaterially shining on me and seized hold of my whole mind and soul, so that I was struck with amazement at the unexpected marvel and I was, as it were, in ecstasy.[44]

On this occasion he is filled with grief at the ending of this vision, so wholly absorbing is it of his 'mind and soul', in other words, his whole self, not just his body. His *Hymns of Divine Love* refer repeatedly to such occasions in language even more fervent that that of the homilies. Whether they repeat the one occasion or are a regular occurrence is not spelled out. But what is clear is that there is a cycle here, of repentance and yearning for incorporation with God, leading to a vision of divine light, leading to greater faith and a further vision.[45] This sense of being simultaneously in the present and seeking future perfection replicates what Symeon teaches about the power of penitent grief, which combines both sorrow at the current distance placed between the penitent and God, and joy that renouncing the sin will bring God closer, bring the penitent sinner back to the first harmony of the created order. As Krivocheine suggests, one cannot receive the medicine unless one first goes to the physician.[46] The truly penitent, who is blessed from the earliest promptings of *metanoia* with light, experiences even more light as they progress spiritually. This soteriological insight is fundamental to Eastern Christian teaching and not unique to Symeon. However, his choice of images and the particular focus he places on the charismatic nature of forgiveness is striking.

Allied to this is the focus on repentance itself as not an initial step along the way but 'the sole gateway of all human experience of God', as McGuckin puts it.[47] In the *Fifth Ethical Discourse* he asserts: 'There is no

God. See McGuckin, 'Symeon the New Theologian's Hymns of Divine Eros', p. 189.

[43] Alfeyev, *St Symeon the New Theologian*, p. 226 n. 93.
[44] *Discourses*, p. 200.
[45] *Catechesis* 2 (15), *Discourses*, p. 58–59.
[46] Krivocheine, *In The Light of Christ*, p. 77.
[47] McGuckin, 'Symeon the New Theologian's Hymns of Divine Eros', p. 188.

other way for anyone to know about God unless it is by means of the contemplation of the light which is sent by him'.[48] This reflects the teaching of Maximos Confessor, in the first *Century of Love*, 2. 6, which describes the marker of pure prayer as being that:

> at the very onset of prayer the mind is taken hold of by the divine and infinite light and is conscious neither of itself nor of any other being whatever except of him who through love brings about such brightness in it.[49]

There are several key points in these visions of light. First, light is ubiquitous as a signifier of salvation and expresses his own personal journey, from which he draws a theological framework and doctrinal advice for his monks.

Secondly, there is the crucial implication that this light is intelligible;[50] whilst it *must* be experienced, this empirical event is no merely sensory distraction. The light is perceived by the *nous* and not just the physical eyes. You must see with your 'intellectual eyes', he insisted in his *Fourth Ethical Discourse*.[51] The intellect on its own has no light but is affected by uncreated light which is invisible to bodily eyes.[52] Golitzin explains Symeon's religious anthropology clearly:

> Man was created 'double', i.e. as body and created spirit, in order for his 'spiritual senses' to be filled with the light of the 'sun' of heaven as the physical eye receives the light of the sun which shines on earth.[53]

In other words, an integration of physical and supraphysical modes of being takes place, such that the whole human person is elevated to God by an ecstatic experience of uncreated light. Numerous references to this occur in the homilies[54] and this resonates clearly with biblical instances of

---

[48] *Fifth Ethical Discourse*, in *On the Mystical Life*, trans. A. Golitzin, vol. II, p. 52.

[49] Maximus Confessor, *Selected Writings*, trans. G. Berthold (New York: Paulist Press, 1985), p. 47.

[50] 'Come, then, you who have the intelligible light in yourselves, let us give glory through it to the Father, the Son, and the Holy Spirit', *Catechesis* 28 (13), *Discourses*, p. 307.

[51] Alfeyev, 'The Patristic Background', p. 237.

[52] Alfeyev, *St Symeon the New Theologian*, p. 236.

[53] Alfeyev, 'The Patristic Background', p. 238, quoting from *Hymn* 23, ll. 448–475.

[54] 'He is [...] the Maker of Light, and the Lord of life. He is the Light that is ineffable, inaccessible', *Catechesis* 19 (6), *Discourses*, p. 230; see also *Catechesis* 28 (13), *Discourses*, p. 307; and *Catechesis* 35 (9), *Discourses*, p. 365: 'Yet God does not show Himself on a particular pattern or likeness, but in simplicity, and takes the

divine light as an indication of the presence of God, forming a sort of spiritual synasthesia. The primary biblical paradigm would be the transfiguration, the uncreated light into which the figure of Christ was transformed as he encountered God.[55] The association of light with Paul's conversion has already been noted, and patristic commentators also claim that Anthony the Great's hearing of the voice of Christ as being accompanied by 'the light descending unto him.'[56] This sense of receiving God as light is stressed as having a Trinitarian aspect, which recalls the line taken by Symeon in the *Third Theological Discourse*. The key point here is that the light of which Symeon is talking is described as intelligible, not merely something perceived by the senses. Its intelligibility makes it infinitely superior to the light of the senses, graphically defended in the *Second Theological Chapter*:

> The sensible sun rises to shine on the world of sense and all it contains [...] but then it goes down and again leaves in darkness the place where once it shone. The intelligible [sun] shines eternally, and was shining, complete in all complete reality, yet not contained by it.[57]

Again and again Symeon stresses the distinction between what he calls the 'sensible' and the 'intelligible': the light affects first the intellect, the *nous*, and only secondarily the body.[58] In *Hymn* 35, he again spells this out in language reminiscent of Ephrem the Syrian:

---

form of an incomprehensible, inaccessible and formless light'; cf. *Catechesis* 20 (6), *Discourses*, p. 236: 'Your mind will see him in the form of a spiritual light with deep calm and joy. This light is the prelude of the eternal and primordial light; it is the reflected brightness of everlasting blessedness.'

[55] Note also *Hymn* 50, lines 31–43: a 'spherical light, gentle, divine, without form, without contour, in a form without form.'

[56] Athanasius, *Vita* 10.1, quoted in Alfeyev, 'The Patristic Background', p. 234, and p. 232, cf. *Catechesis* 6 (2), *Discourses*, p. 120.

[57] Section 2. 22, in Symeon, *The Practical and Theological Chapters*, trans. P. (=J. A.) McGuckin, pp. 70–1, cf. *Catechesis* 28 (10), *Discourses*, p. 303: 'The visible sun gives light only to physical eyes, and not of men alone, but also of irrational beasts, quadrupeds, and birds; the intellectual sun, however, that has appeared in the world, gives light to rational souls only. Those it does not enlighten indiscriminately, without their being worthy.'

[58] Alfeyev, 'The Patristic Background', p. 236; cf. Alfeyev, *Symeon The New Theologian*, p. 235.

Sensible lights illumine only the bodily eyes; they shed light and make visible the sensible but not the intelligible realities [...] the spiritual eyes of the heart must therefore also be enlightened by a spiritual light.[59]

Thirdly, Symeon stressed that being a demonstrably illuminated person is what confers power to remit sin, over and above ecclesiastical status, as proved by the authority he attributes to Symeon the Studite.[60] This reinterpretation of conventional channels of authority brought Symeon into such conflict with the church of his day that he was exiled and accused of fostering a cult of his spiritual father, whose association with visions of light had so strongly shaped his transition from the Byzantine court into the monastic cell. The yoking together of divine illumination and repentance is beautifully developed in the *Catechesis* 23, but the whole issue of individual repentance and salvation is at the heart of all Symeon's writings.

This brings us finally to the image of sunlight passing through a glass of wine. *Catechesis* 23, written for his newly aquired monks at the monastery of St Mamas in Constantinople, pulls together two of Symeon's favourite themes – repentance and light mysticism – to teach his rebellious monks that true knowledge of God is not just the *telos* of the spiritual journey, it is its essential starting point. Many traditional understandings of the Christian progression followed the Evagrian model of *praxis*, *theoria* and *gnosis*, a taxonomy expressed throughout eastern and western Christendom in various tripartite guises. Symeon, however, asserted that illumination is the start as well as the culmination of repentance, which itself forms the vehicle for union with God. 'Knowledge', he writes, 'is not the light! Rather, it is the light that is knowledge, since "in it and through it and from it are all things" (Romans 11. 36)'.[61] It is as if he follows the Aristotelian rather than Platonic understanding of the source of knowledge, which, according to Adam Becker is: 'perception (as opposed to Plato, who puts the intellect first)'.[62] True knowledge, accompanied by tears, discloses the divine Light, and light constitutes knowledge.[63]

---

[59] Lines 58–64, quoted in Krivocheine, *In the Light of Christ*, p. 231.
[60] *Catechesis* 28 (6), *Discourses*, p. 299, *Catechesis* 33 (2), *Discourses*, p. 340.
[61] *Catechesis* 28 (7), *Discourses*, p. 301.
[62] A. H. Becker, *Fear of God and the Beginning of Wisdom* (Philadelphia, PA: University of Pennsylvania Press, 2006), p. 148.
[63] *Catechesis* 2 (9), *Discourses*, p. 54.

Similarly he states that repentance, expressed by joyful tears (penthos), should accompany all stages of the process of enlightenment. He thus posited a sort of realised eschatology, redemption both now and to come. This fitted with his view of salvation as a restoration to the perfection of Eden, a stripping away of the layers of sin which dull and cloud the image of God in humanity. Symeon inherited a patristic anthropology which sees Christ as recapitulation, and the human being as capable of restoration and perfection. The human nature he envisaged being perfected is both physical and spiritual, a human image of the divine nature of Christ, which—like Christ—combines inextricably apparently contradictory qualities. This provides the perfect setting for an image of repentance which mingles earthly and heavenly experiences. Such radical and disturbing demands for insight and wisdom preceding spiritual advancement must be placed in the context of Symeon's own life and interpretation of mystical paradigms.

The image featured in this homily involves not only sunlight but wine, and this merits some consideration. Wine was, after water, the staple drink of the Byzantine, both layperson and monk, as indicated in *typika* of the day which record suggested amounts for daily consumption, starting at breakfast.[64] As well as providing essential fluids it had medical and culinary uses and its place within monastic communities was even defended by Basil.[65] The symbolic significance of wine within the Eucharist is paradigmatic, but there are other clear Biblical examples of the cultural practices surrounding wine which perhaps underlie Symeon's image of the winepress in *Catechesis* 23. The fumigation of wine vats with incense before use suggests not only a purifying process, but perhaps also a religious connotation. Also, wine formed a crucial part of the Byzantine economy, not least for those monasteries which traded in it, rather than exchanging it.[66] Perhaps some local experience of wine production at St Mamas inspired Symeon when detailing the action of the winepress: typically, he drew on a commonplace example as a means of teaching. Speaking of the connection between suffering and joy, and how repentance leads to bliss, Symeon described how the Holy Spirit 'presses and filters his heart' as if through the action of winepress crushing grapes. The result is joy whose purity is stressed by comparing it to wine that has been filtered, and is then raised in front of the sunlight. Both the

[64] *Oxford Dictionary of Byzantium*, III, 2199.
[65] Ep. 199.47.10
[66] *Oxford Dictionary of Byzantium*, III, 2200.

luminosity and colour of the wine is stressed, as it 'flash[es] joyfully on the face of him who drinks it as he faces the sun'.[67] The sense of joy and light is described as a gracious gift, the reward of the Holy Spirit for penitence and 'bitterness [...] of soul'.[68]

To place this image in context, it is helpful to have a basic outline of *Catechesis* 23. The homily skilfully weaves images of sin as ingesting poison, and repentance as health-giving wine. It begins in the third person, describing the anguish of the sinner in terms of physical discomfort. Symeon has a remarkable ability to evoke the sensual and corporeal as very much part of the human condition. The second section moves to a direct appeal by the sinner to God for mercy. In acknowledgement of his repentance, God's mercy is literally poured out:

> He will change the bitterness of his heart into the sweetness of wine and will cause him to spew forth the poison of the dragon that was burning up his innards.[69]

Pain has become joy, but only through the workings of the Holy Spirit which has acted on him like a winepress—again another very physical image, to which he returns by the end of the homily, urging the sinner to repent and embrace the fear of the Lord which 'wastes the flesh and breaks the bones, just as the stone moved by the mechanism presses the grapes that are in the winepress and crushes them completely.'[70]

The resulting wine of repentance is 'held up against the sunlight' like a chalice at the Eucharist, a holy offering of sincere repentance which itself further illuminates the penitent, and confers immortality and joy. The image is sensuous, evoking intense beauty and hope:

> As the Spirit presses and filters his heart [as in a winepress], so it produces a joy that is genuine and unmixed with affliction. For this reason death will have no dominion over it – no blemish will be found in it. But it will be like wine that has been strained and is held up against the sun[light] shining brilliantly and showing its colour more clearly and flashing joyfully on the face of him who drinks it as he faces the sun.[71]

At this point, Symeon adopts an almost cosy familiarity with his reader, describing in the first person the immediacy of his sensations and the

---

[67] *Catechesis* 23, *Discourses*, pp. 257–58.
[68] *Catechesis* 23, *Discourses*, p. 257.
[69] ll. 107–9, *Discourses*, p. 256.
[70] ll. 206–9, *Discourses*, p. 259.
[71] *Catechesis* 23 (4), *Discourses*, p. 257.

process of attempting to rationalise them into a theological framework. The taste, the sight and the sheer ascetics of the sunlight shining through the wine are evoked, as he describes an insatiability for the life-giving repentance:

> But in these matters there is one thing that I cannot understand. I do not know which pleases me the more, the sight of the sun's rays and the delight of their purity, or the drinking and the taste of the wine in my mouth. I would say it is the latter, yet the former attracts me and appears more pleasant to me. Yet as I look at it, I derive more pleasure from the sweetness of tasting, so that I am not sated with seeing nor filled from drinking. For when I think that I have drunk my fill, then the beauty of the rays that pass through it redoubles my thirst and I crave it again. The more I am eager to fill my stomach, my mouth burns ten times as much and I am inflamed by the thirst and desire for that most transparent drink.[72]

Stepping back a little from the intensity of the experience, he switches into the third person to affirm the inseparability of the light and the taste, and how both physically and spiritually they transform and empower the whole person of the one who 'drinks' repentance in this way.

> The shining of the wine and the beam of the sun as they shine on the face of him who drinks penetrate to his inward parts, to his hands, his feet, his back, and transform him wholly into fire. They give him the power to burn and melt the enemies that approach him from every side. He becomes dear to the sunlight and a friend of the sun. To the transparent wine belonging to the rays that issue from it he becomes like a beloved son, for the drink is his nourishment purging the infection of his putrified flesh.[73]

The experience of God as light is so profoundly desirable that the penitent yearned continually to remain in this state of sober inebriation. The image of addiction to the sunlight recalls an image of fire enticing the penitent in a vision of divine incorporation. In *Hymn* 25, Symeon described encountering God as a column of light, which prompts him to a fervent prayer of repentance invoking the Trinity : 'Oh! Ebriety of the Light! Oh risings of fire! Oh the flame's swirlings moving in me, the wretched one.'[74] The fire, the sunlight, can hardly be separated from the person of the ecstatic young monk. Theologically, the indivisibility of the heat of the fire (or light of the sun) from the perishable body of the human mirrors the inextricably meshed humanity and divinity in Christ,

---

[72] *Catechesis* 23 (5), *Discourses*, p. 258.

[73] *Catechesis* 23 (5), *Discourses*, p. 258.

[74] ll. 23–39, quoted in Krivocheine, *In the Light of Christ*, p. 230.

or the perichoresis of the Trinity. And the apparent inability to 'understand' which is better, the sun or the wine, insists on an apophatic reading of the vision of God: it goes beyond words and reason into sensation and instinct. This complete integration of the divine empowers the one who experiences it. Having directly experienced God's love and forgiveness through the immediacy of the non-verbal, uncreated light grants the charism to transmit forgiveness to others. Symeon insisted that if you do not have the light within you, then you are 'unable either to lead others or to teach them the will of God', nor are you 'fit to hear (in confession) the thoughts of others'.[75] Cleverly, he reinforced this by returning to the image of fear of God and repentance as the mystical winepress, in a direct address to his 'beloved children'. Since light was often read by him as being the presence of God as Trinity, he was calling on the very highest authority to back his worldly status as Abbot.[76]

It has already been noted that penitence is both the gateway and the door which is luminous in itself, as well as leading from the darkness into light. He who does not enter into the light (cf. John 3. 20) has not properly passed through the gate of repentance, for had he done so, he would have been in the light. Given that one cannot separate the light from the colour when holding up a glass of wine in the sunlight, this whole homily suggests a human encounter through repentance with the dual nature of Christ, and the undivided Trinity.

---

[75] *Catechesis* 33, *Discourses*, p. 340.
[76] Krivocheine, *In the Light of Christ*, p. 237.

LUX MEDIATRIX: 'ALL THAT IS MADE MANIFEST IS LIGHT' (EPHESIANS 5. 13):
THE INFLUENCE OF ROBERT GROSSETESTE'S THEORY OF LIGHT ON BISHOP JOHN FISHER.

Cecilia A. Hatt

During the first decade of the sixteenth century, probably in 1507, John Fisher, the bishop of Rochester, preached a series of sermons on the Penitential Psalms to the household of Lady Margaret Beaufort, to whom he was chaplain. On 8 September, the feast of the nativity of the Virgin Mary, he was due to preach on the psalm *Domine, ne in furore tuo arguas me, neque in ira tua corripias me* (Rebuke me not, O Lord, in thy indignation; nor chastise me in thy wrath).[1] The bishop begins his sermon, rather disarmingly, in this way:

> Meruayle no thynge all though we begynne not our sermon with the thyrde penitencyal psalme in ordre. For or euer we toke vpon vs to declare the two fyrst penytencyal psalmes our promyse was som what to speke of the natyuyte of our blessyd lady at the daye/ whiche purpose wyllynge to kepe/ also desyred of our frendes to folowe thordre of the psalmes/ though it semed to be harde for vs so to doo. Notwithstondynge by the helpe of our blessyd lady we haue attempted the mater and made the fyrst parte of this psalme to agre with our fyrst purpose.[2]

The sermon that follows offers an elaborate example of an extended prelocution based on a secondary text, a structural device frequently used in the university sermon, but not always achieved so well as it is here. Fisher takes his secondary text from the text of the day,

> *Quae est ista quae progreditur quasi aurora consurgens* [Who is she that cometh forth as the morning rising][3]

---

[1] Psalm 37/8. Except for those provided by Fisher himself, English translations of Scripture are taken from the Douay-Rheims version.
[2] *The English Works of John Fisher*, ed. J. E. B.Mayor, EETS, e.s. 27 (1876), pp. 44–51 (p. 44).
[3] Song of Solomon 6. 9.

After the offence of our fyrst faders Adam and Eue/ all the worlde was confounded many yeres by derkenes and the nyght of synne of the whiche derknes and nyght a remembraunce is made in holy scrypture often tymes. Notwythstondynge many that were the very seruauntes and worshyppers of almyghty God to whome the sayd derkenes and nyght of synne was very yrksome and greuous/ had monycyon that the very sonne of ryghtwysnes sholde sprynge vpon all the worlde and shyne to theyr grete and synguler comforte and make a meruaylous clere daye. As the prophete Zacharye sayd and prophecyed of Cryste. *Visitauit nos oriens ex alto/ illuminare his qui in tenebris et in vmbra mortis sedent.* Our blessyd lorde hath visyted vs from aboue to gyue lyght vnto them whiche syt in derkenes and in the shadowe of deth. [...] Neuertheles theyr good hope and trust of it was dyfferred many yeres/ and at the last whan tyme was houable and conuenyent in the syght of almyghty God/ he caused this clere sonne to gyue lyght vnto the worlde. Notwithstondynge it was done in a Iuste and due ordre. For of a trouth it had not ben semynge and well ordred that after soo grete and horryble derkenes of the nyght/ the meruayllous clerenes of this sonne sholde haue ben shewed immedyatly. It was accordynge of very ryght that fyrst a mornynge sholde come bytwene whiche was not soo derke as the nyght neyther so clere as the sonne. This ordre agreeth bothe to nature/ scrypture/ and reason. Fyrst by the ordre of nature we perceyue that bytwene the derkenes of the nyght and the clere lyght of the daye/ a certayne meane lyght cometh bytwene whiche we call the mornynge/ it is more lyght and clere than is the nyght/ all be it the sonne is moche more clerer than it. Euery man knoweth this thynge well/ for dayly we haue it in experyence.¶ Holy scrypture also techeth that in the begynnynge of the worlde whan heuen and erthe sholde be create/ all thynges were couered with derkenes a longe season/ and or euer the sonne in his very clerenes gaue lyght to the worlde/ a certayne meane lyght was made whiche had place bytwene derkenes and the very clere lyght of the sonne. This is well shewed by Moyses in the begynnynge of Genesis. ¶Reason also whiche sercheth the knowlege of many causes fyndeth whan one thynge is chaunged into his contrary as from colde to hete/ it is done fyrst by certayne meanes or by certayne alteracyons comynge bytwene. ¶ Water whiche of his nature is very colde is not sodeynly by the fyre made hote to the vttermost/ but fyrst cometh bytwene a lytell warmenes as we myght saye lukewarme/ whiche is neyther very hote nor very colde/ but in a meane bytwene bothe. ¶ An apple also whiche fyrst is grene waxeth not sodeynly yelowe/ but fyrst it is somwhat whyte bytwene grene and yelowe indyfferent. Thus we perceyue by reason that it was not conuenyent this grete clerenes of the sonne our sauyour sholde haue ben shewed soo soone and immedyatly after soo ferefull and the derke nyght of synne/ without rysynge of the mornynge whiche is a meane bytwene bothe. [...] We se by experyence the mornynge ryseth out of derknes as the wyse man sayth. *Deus qui dixit te tenebris splendescere.* Almyghty God commaundeth lyght to shyne out of derkenes. The clerke Orpheus meruayleth gretely of it sayenge. *O nox quae lucem emittis.* O derke nyght I meruayle that thou bryngest forth lyght. And of a trouth it is meruayle to mannes reason that lyght sholde sprynge out of derkenes. Laste all though it semeth the mornynge to be cause of the sonne/ notwithstondynge the sonne without doubte is cause of it. And in lyke wise although this blessyd vyrgyn brought forth our sauyour Ihesu/ yet he made her and was cause of her bryngynge in to this worlde. Thus ye perceyue by nature that this blessyd vyrgyn may well be lykened to a mornynge [...] [4]

Two things are immediately noteworthy. The first is Fisher's recourse to the concept of the *mean*, the interim stage between two states; and the second is his appeal to the three authorities, nature, scripture and reason, that all witness to the universal significance of this concept. After the prelocution, bishop Fisher returns to his psalm. He establishes three especial qualities attributable to the Virgin, mild restfulness, a disposition to move towards the light, and clarity, and he sets these three qualities against the three tribulations that are summed up in the first half of the psalm: fear, bondage and ignorance.

> Cor meum conturbatum est; dereliquit me virtus mea, et lumen oculorum meorum, et ipsum non est mecum.[5]
>
> (My heart is troubled, my strength hath left me, and the light of my eyes itself is not with me.)

This sermon deals only with the first part of the psalm, and Fisher's use of the morning-theme is worth closer consideration. In order to see how his approach is significant it is useful to compare it with a passage that is superficially similar. Peter Dronke quotes the lyric *In rosa vernat lilium*, in which the sun and moon are used as figures of God and the Virgin Mary:

> Ex luna solis emicat
> radius elucescens:
> mundanis solem indicat
> luna nunquam decrescens.
> Hic sol dum lune iungitur,
> neuter eclypsim patitur,
> sed est plus quam nitescens.

---

[4] *English Works*, pp. 44–48. Douglas Gray prints part of this prelocution in his *Oxford Book of Late Medieval Verse and Prose* (Oxford, 1985), pp. 117–19.
[5] Psalm 37. 11.

> From the moon shines forth the dawning ray of the sun, the moon that never wanes shows the sun to mankind. When this sun is united to its moon, neither suffers eclipse, but each is more than radiant.[6]

This is a neat analogy. The moon has no natural light of its own but shines with the reflected light from the sun, which is theologically acceptable, although the implications of the last line are not: the moon cannot give to the sun any more light than it already has. However, it is clear that the poem is using a metaphor and although Dronke's discussion of this figure demonstrates how sophisticated it can become, it is still essentially a figure, a conceit. Fisher, it appears, is expressing something different. For one thing, the Virgin is not linked to a heavenly body in what one might call a decorative or complimentary way, but to a process. As he says, quite specifically, he is not primarily occupied in finding praises for the Virgin

> our mynde at this tyme is not to speke of her laudes whiche no creature can sufficyently expresse [...][7]

What he is doing is to establish her as implicated in a daily, wonderful, but entirely natural, phenomenon. The theory of the 'interim light' recalls various interpretations of Genesis. The first three days of creation, when light had been called forth, but the heavenly bodies were yet to appear, had been discussed at length by the Fathers.

Augustine thought of it as a kind of intellectual or spiritual creation and, because there could never have been an absence of uncreated light Augustine formulated a theory about a 'morning knowledge' assigned to angels, or possibly creatures in a state of grace, and 'evening knowledge' assigned to others.[8] The ps. Bede also atttributed an angelic quality to the first light.[9] Aquinas too, touched on this topic, but his and those of the earlier theologians were fundamentally allegorical interpretations of the scripture.[10] A slightly different understanding of the interim light can be found in the work of Robert Grosseteste, who, in his short work, *De luce*,

---

[6] *Medieval Latin and the Rise of European Love-Lyric*, 2nd edn, 2 vols (Oxford, 1968), pp. 128ff.
[7] *English Works*, p. 52.
[8] See *De Gen. ad litt.* ch. 30 PL 34, col. 316. The correspondence of *light* with *cognition* in this passage is clear.
[9] *De creatione vi dierum*, PL 93, col. 213.
[10] See, for example, *Super Sent.*, lib. 2 d. 2 q. 1 a. 3 arg. 3; id. lib. 2 d. 12 q. 1 a. 3 co., and passim.

set out the theory, original to him, that light is the first corporeal form and effective principle of movement. He begins by using the terminology of Aristotle but not necessarily the ideas:

> [his] chief point of divergence is that for Grosseteste matter is not pure potency, as it was for Aristotle, but possesses in its own right a certain minimal reality. Thus Grosseteste speaks of matter as a substance [...][11]

Grosseteste describes form as actualising matter, in the sense of completing it—he uses the terms *compleo* and *complementum*—and the characteristic feature of matter is the capacity of extension, which is notably a property of light.[12] Fisher was familiar with Grosseteste's work (he quotes him several times) and it would seem that he found the tenor of Grosseteste's treatment of light particularly congenial. Fisher's 1521 Paul's Cross sermon draws extensively on Grosseteste's *De calore solis* and *De natura locorum*. The bishop of Rochester describes the effect of the sun's rays on a tree in spring, drawing analogies between the operations of light and of the Holy Spirit. The sermon was preached on 12 May 1521, and he remarks:

> What meruaylous vertue, what wonderfull operacyon is in the bemes of the sonne whiche, as we se this tyme of the yere, spred vpon the grounde dothe quycken and make lyfely many creatures the whiche before appered as deed! Who that vewed and beheld in the wynter season the trees whan they be wydred and theyr leues shaken from them and all the moystour shronke in to the rote and no lust of grenenes nor of lyfe appereth outwardly, yf he had had none experyence of this mater before he wold thynke it an vnlyke thyng that the same trees sholde reuyue agayn and be so lustely cladde with leues and floures as we now se them. And yet this is done by the subtyll operacyon and secrete workynge of the sonne bemes spred vpon the grounde.[13]

He goes on to be more specific:

> It is a trouthe, the bemes of the sonne in wynter be lyght as they be now this tyme of the yere, but that lyght is so faynt and feble that it gyueth no lyfe, for than we sholde haue herbes and trees to growe as well in wynter as they now do this tyme of the yere. The cause of this weykenes is: for the sonne shooreth so lowe by the grounde that his bemes than sklaunteth vpon the grounde and dothe not rebounde nor double in theymselfe agayne towardes the sonne, and this is the cause of this weykenes. Ye se whan a bowle is

---

[11] *De luce,* trans. Clare Riedl (Milwaukee,WS, 1942), p. 3.

[12] Ibid. p. 5.

[13] 'Sermon made agayn the pernicious doctryn of Martin luther', in *English Works of John Fisher,* ed. Cecilia A. Hatt (Oxford University Press: Oxford 2002), p. 83.

throwen sklantlynge vpon a wall it slydeth forwarde and reboundeth not bakwarde dyrectly agaynst to hym that was the thrower, but whan it is dyrectly cast agaynst a wall with a grete vyolence than it dothe dyrectly rebounde agayne. [the angle of incidence is equal to the angle of reflection] In this maner it is of the sonne bemes: the more nye that the sonne draweth vnto vs now this tyme of the yere, the more dyrectly his bemes bete vpon the grounde and the more dyrectly they rebounde and retourne agayne towardes the sonne. And by the reason of the nyenesse of beme to beme ryseth a greter strengthe in the beme and a more full lyght.[14]

Grosseteste's account of seasonal variations is clearly influential:

> And, the further the place is from the equinoctial, the more obtuse the angle at which the sun's rays strike and are reflected, and by so much less do the incident and reflected rays shine on the opposite plane, so the division is less and less heat is generated.[15]

The passage in this Paul's Cross sermon is peculiarly characteristic of Fisher. On the most simple level it shows him observing with delight the beauties of nature, something which is a feature of all his English writing. More than that, he is deeply satisfied by the ontological coherence of the whole phenomenon. This how he interprets what is going on:

> This example yf ye perceyue it maye enduce vs to conceyue how wonderfully the spyrytuall sonne almyghty God worketh by his spyrytuall and inuysyble bemes of his lyght spred vpon the soule of man or vpon the chyrch, bothe whiche is called in scrypture a spyrytuall erthe. *Dominus dabit benignitatem et terra nostra dabit fructum suum.* That is to saye, Our lorde shall gyue his gracyous influence and our erthe shal yelde fruytfull workes. The bemes of almyghty God spred vpon our soules quyckeneth them and causeth this lyfe in vs and the fruyte of good workes. Fyrst they cause the lyght of faythe, but this is a veray sklender lyght withouten the reboundynge of hope and the hete of charyte. Faythe withouten hope is a sklender beme and of a lytle power. But Ioyne vnto hym hope whiche reboundeth vp to God agayne *ad ea quae non videntur* and than is he moche stronger than he was byfore. For

---

[14] Ibid., 83–84.

[15] Et quanto magis distans fuerit locus ab aequinoctiali, tanto magis cadunt et reflectuntur radii solares secundum angulos obtusiores et tanto minus in partes oppositas cedunt radius cadens et radius reflexus et minor fit disgregatio et minus generatur calidum. *De calore solis, Die Philosphischen Werke des Robert Grosseteste,* ed. L. Baur, Münster 1912, *Beiträge zur Geschichte der Philosophie des Mittelalters* Bd. IX, 83.

> nowe this is doubled and boughted in it selfe and gadred more nye vnto himselfe and made more valyaunt and mighty than it was before.[16]

Clearly, Fisher is speaking analogically here: he is not saying that the beams of light that shine directly onto a wall are, so to speak, particles, or waves, of faith or of hope. However, he is saying that the virtues operate according to exactly the same principles as beams of light and in this he was agreeing with Grosseteste, one of whose favourite ideas was that of the 'congregation' or multiplication of forces, an idea with obvious didactic potential. Throughout Bishop Fisher's work, light has a special significance, not only as a metaphor for the knowledge and love of God but as an actual manifestation of it. Darkness and tempest are thus, for him, also objective signs of ontological disorder. Grosseteste also, who, while he was ready to remark the analogical or metaphorical functions of light, like the ps. Denys or other Christian Neoplatonists, insisted on its literal corporeity, as the active principle in everything.

> Light [...] is that by which the soul acts in all the senses, and is that which acts as an instrument in all of them.[17]

It must be so, according to Grosseteste, because light is the substance most capable of diffusing itself outwards. Light always tends to spread into a sphere around its source and bears a greater resemblance to the intelligences than any other extra-material form. Because of its infinite facility in being reflected or refracted, light is the source of all matter. Grosseteste proves this by arguing that a simple entity, such as light (i.e. with no dimensions but position), multiplied an infinite number of times, will produce a finite number of dimensions:

> It is clear that every higher body, in virtue of the light which proceeds from it, is the form and perfection of the body that comes after it. And just as unity is potentially every number that comes after it, so the first body, through the multiplication of its light, is every body that comes after it.[18]

In the *Hexaemeron*, chapter X of which paraphrases the opening of *De luce*, Grosseteste writes of the six days of creation, to which created light is central, because it is both the enabler and the form of it:

---

[16] Hatt, *Works*, p. 84.
[17] *Hexaemeron*, trans. C. F. J. Martin (Oxford: Published for the British Academy by Oxford University Press, 1996), p. 98.
[18] *De luce*, p. 15.

The light which is established, then, means in the first sense the visible light that went on through the first three days in a temporal way. In the second sense it means the nature of the angels, turned back to the contemplation of God. In the third sense the establishment of light can be taken to mean the bringing of formless matter to forming. For every form is a kind of light and the manifestation of the matter that it forms, as St. Paul says 'all that is made manifest is light'.[19]

This first light then, that preceded the creation on the fourth day of the heavenly bodies, is a sort of epistemological substrate—that which would be knowable when creation had gained the form to receive it. The sermon on the penitential psalm shows how Fisher finds this interim light particularly referable to the role of the Virgin Mary because of her unique position as having, so to speak, a foot in both camps:

> lyke as the mornynge is a meane bytwene the grete clerenes of the sonne and the vgsome derkenes of the nyght. So this blessyd and holy virgyn is the meane bytwene this bryght sonne our sauyour and wycked synners/ and a partetaker of bothe/ for she is the moder of Goddes sone and also the moder of synners [...] Saynt Austyne sayth it semeth to be a noble kynrede bytwene this blessyd vyrgyn and synners/ for she receyued all her goodnes for synners/ synne was cause why she was made the moder of God. Also yf we haue taken ony goodnes we haue it all by her. Therfore of very ryght this holy virgyn Marye is the moder of synners.[20]

It is as if she is placed at the point where potency becomes act. Sin is, as so often, behovely; Mary's necessary involvement with sinful humanity is, for Fisher, a linkage with active force, not just a 'mean' in the sense of something that happens to be located halfway between two things. Mary is both a means towards enlightenment and by grace of her virtues a source of light in herself. John Fisher is quite clear that it would have been somehow unfitting for the light of Christ to be suddenly apparent:

> it had not ben semynge and well ordred that after soo grete and horryble derkenes of the nyght/ the meruayllous clerenes of this sonne sholde haue ben shewed immedyatly[21]

Just as no one becomes all of a sudden most base, neither does anyone become all of a sudden most enlightened. The road to Damascus, the 'all at a crash Paul' experience, occasionally happens as an extraordinary

---

[19] *Hexaemeron*, p. 96.
[20] *English Works*, pp. 49–50.
[21] *English Works*, p. 46.

grace, but what Hopkins called 'the lingering out sweet skill' of conversions like Augustine's is closer to the general rule. This is because it *is* 'semynge and well ordred' that human beings should learn from each other and from the world about them. Light, whether understood intellectually or corporeally, illuminates by being reflected, and

> this blessyd lady Marye as a mornynge goth bytwene our nyght and the day of cryste/ bytwene our derknes and his bryghtnes/ and laste bytwene the mysery of our synnes and the mercy of God[22]

The process implied here goes both ways; human beings learn from the light that is reflected from God and from other people. The image therefore is of constant and generous replenishment of the world; the will is God's but the love which accompanies it seems to bounce back and forth like the light that embodies it. This idea of the knowing co-operation between created things, made possible by the fact that in the first light the principles of knowing and of being coincide, is surely what lies behind Fisher's evocation of the morning, and it almost certainly owes a great deal to Grosseteste, who does not propound a general pantheistic consciousness, but a tendency specific to each individual thing to reflect its being outward to the extent of the light that is in it. This is not to say that either Grosseteste, or Fisher in quoting him, was unaware of the possibility of using physical phenomena simply as metaphors. Grosseteste was comfortable, for example, with the metaphor of shadows or veils which was such an important part of the rationale behind the use of stained glass in twelfth-century architecture. However, it is perhaps not helpful or relevant to attempt a classification of language as tropical or not by suggesting that talk about the phenomenon of light is literal at the secular end of the continuum and metaphorical at the sacred.

For Grosseteste it was a simple scientific fact that the physical universe is made up of light. That this was so was an idea of Plotinus', but Grosseteste's view that light is a substance is a considerable advance on it, and marks a step from metaphysics to physics that was truly original. This mingling of theology with natural science came to be a distinguishing mark of what was known as the 'Oxford school'. Grosseteste's studies were solidly founded in physics; whatever he might have thought in his speculative moments about the dialectic of being and non-being, in his writings, he was entirely concerned with the physical fact of being and the dynamism of growth and change. These result, he said, from the genera-

[22] Ibid., p. 51.

tive power of light acting upon matter. A couple of centuries later we see Fisher taking a step the other way, from physics to metaphysics:

> whan the bemes of faythe and hope be ioyned togyder in one poynt than it is of myghty power. The bemes of the sonne whan by reflexyon of a brennyng glasse they be gadred togyder, they be so myghty that they wyll set tynder or clothe on fyre. And lykewyse it is of the bemes of the faythe and hope whan they be ioyntly compacte and vnyte togyder.[23]

It does not seem at all fortuitous that the scientific enquiries of scholars such as Grossesteste and Albertus Magnus from which John Fisher was drawing lessons about knowledge are enquiries about the act of reflection. For Grosseteste, unlike Aristotle, this was not a passive affair. He believed in the emission of rays: that light went out from the eye and mingled with rays coming in:

> Light is therefore the instrument of the soul in sensing through the senses of the body.[24]

That which is created, then, takes a part in its own recreation and extension: each creature, being a form of light, reflects the infinite light in its own finite way. Just as Fisher perceives the creative operation of the Trinity in the election of Mary:

> this blessyd virgyn was ordeyned by the hole trynyte to sprynge and to be brought forth in to the worlde [25]

so had Grosseteste used the principle of uncreated light as a way of looking at the Trinity:

> The presence all together in visible light of three factors, light, splendour and heat (which are all one in essence but distinguishable by thought) offers us a model (*demonstratio per exemplum*) [...] for thinking the Trinity of God.[26]

James McEvoy describes it like this:

> What is reflected back into the primordial, infinite radiation is its very own nature, a nature that has never left it [...] The unity and identity of what is radiated forth, received, and then reflected back by the combined radiating action of source and image, is in turn a third fact or aspect of the infinitely dynamic nature of God [...] the inexhaustible intelligibility of the plenitude

---

[23] Hatt, *Works*, p. 84.
[24] *Hexaemeron*, p. 98.
[25] *English Works*, p. 49.
[26] Cf. *Hexaemeron*, pp. 226–27.

of being [... is] at once the action of manifesting [...] and the state of its being manifested.[27]

The act of creation thus is also intrinsically an act of reflection, whereby God's calling forth of the light is performed with reference to the creatures who will be created to see it. Augustine had suggested that it was an intellectual light, but Grosseteste elaborates this by asserting that created light became most truly itself once it had reference to created forms. In the *Hexaemeron* he makes the bold claim that

> 'he saw it' means that he made it to be seen by us. For he does what we do in him and what he does in us.[28]

Creation is therefore designedly intelligible to the human mind.

To believe this is to entertain a very particular understanding of the reality-value of a reflection. Augustine, in *The City of God* at any rate, maintains a Platonist scepticism about the need to pay much attention to sciences that are concerned with visible objects. With the exception of medicine, physical studies are peripheral to his idea of what is really needful for the Christian life. But later Christian writers, influenced by Aristotle, had a very different attitude towards the natural sciences, of whose importance *in themselves* they were profoundly convinced. '*In themselves*' is stressed here because it could be argued that the frequent medieval emphasis on the specular character of physical nature indicates a valuing of such scientific studies only insofar as they threw light on theological truth. One thinks in this context of the theory as set out, for example, in Alain de Lille's *Omnis mundi creatura*, whereby nature as it were holds up a mirror to God, and that image, of the world reflecting the things of Heaven, is of course everywhere to be found. Even so, common as it is, the trope has its own ambiguities, because the ontological status of a mirror image will be different depending on whether it is observed by a Platonist or an Aristotelian. The Platonist reflection is a shadow of a shadow, with a diminishing relationship with reality. Thomas Aquinas on the other hand analyses the process of perception, whereby the sight sees the likeness of the visible thing by means of the intelligible species, but that which is primarily understood is the object of sight itself:

> since the intellect reflects upon itself, by such reflection it understands both its own act of intelligence, and the species by which it understands. Thus the

---

[27] *Robert Grosseteste* (Oxford: Oxford University Press, 2000), p. 93.
[28] *Hexaemeron*, p. 92.

intelligible species is that which is understood secondarily; but that which is primarily understood is the object, of which the species is the likeness[29]

Fisher's understanding of the epistemological role of the Virgin Mary fits rather neatly into this shape: she is in herself, as all human beings may be, a kind of intelligible species, but that which is primarily understood is God himself.

At the beginning of this article attention was drawn to two things, Fisher's concept of the mean, and his appeal to the three authorities, nature, scripture and reason, that all witness to the universal significance of this concept. It is important to point out that, apart from what has been called the decorative quality of light as connected with the Virgin Mary (and in Fisher's work, clear light is always synonymous with beauty), the dawn, the morning that this sermon celebrates refers to an access of *knowledge* that has been and continues to be, facilitated by a system of intermediaries. The intermediaries or means are not necessary evils but themselves objects of celebration, as Fisher demonstrates with his invocation of nature, scripture and reason to support them. Human reason is not something separate from nature or indeed from scripture, but a capacity that gives point and purpose to what we learn from those two books, and more than that, enables us to will the reciprocity that is for other animals only instinctual. Of the patterns in creation identified by natural philosophy, Bishop Fisher here invokes the pattern of enlightenment derived from reflection and familiarity, and that continuing enlightenment, to which the everyday world bears witness, is constitutive of the continuance of creation.

An effective preacher's discourse has to deal with things that change and go away. The value of such things is not negated by reference to the eternal things; instead it is established and put in its place. Fisher's use of ordinary mortal things, of dawn and twilight, hot water and apples, in analogies with the eternal, implies his taking–for–granted of grace in the world. During the main part of his sermon, John Fisher connects Mary's role as mediatrix of divine light to the three wretchednesses of human life, fear, bondage and ignorance. The light of which she is privileged to be a vehicle is an epistemic agent, both object and means of knowledge, and is not really metaphorical at all. He has ingeniously accommodated to each

---

[29] 'quia intellectus supra seipsum reflectitur, secundum eandem reflexionem intelligit et suum intelligere, et speciem qua intelligit. Et sic species intellectiva secundario est id quod intelligitur. Sed id quod intelligitur primo, est res cuius species intelligibilis est similitudo'. *S.T.* I^a q. 85 a. 2 co.

other two seemingly disparate texts and makes no secret of the fact that this took a bit of an effort to achieve:

> though it semed to be harde for vs so to doo. Notwithstondynge by the helpe of our blessyd lady we haue attempted the mater[30]

This is because there is no need to make a secret of it; the world contains such a thing as misery and also such a thing as the morning and therefore they are not foreign to each other. Fisher's sermon is an instance of that which it describes: the process whereby the things of creation are inherently and necessarily bringers of light to each other.

---

[30] *English Works*, p. 44.

SEEING IN SERMONS:
WORD, LIGHT AND AESTHETIC EXPERIENCE

Virginia Langum

Of the various physical traits assigned to the seven deadly sins in *Piers Plowman*, the ocular conditions are among the most vivid and suggestive. Gloton's eyes are 'dymmed', Coveitise's 'blered', Wrath's 'white' and Sloth's 'slymed'. Pride and Envy suffer wayward gazes; Lechery appears to have no eyes at all.

These visual disorders reflect behaviour and moral character. While the spiritual often blends with the physical in physiognomic texts, other encyclopædic works, such as Trevisa's translation of *De proprietatibus rerum*, also associate spiritual conditions with medical diseases, particularly those of the eyes.[1] The condition of the eye and its ability to see takes precedence over other ailments in the body, because disease in the eye is thought to indicate disease in the interior body and soul: 'Amonge alle þe wittes þe yen beþ next to þe soule, for in y3en is þe token of þe soule. For in þe y3en is iknowe and iseye al þe dome of mynde, desturbance and gladnes of þe soule, and also loue and wraþþe and oþir passiouns.'[2] Medical writers also discuss how the interior condition of the body may be known by looking at the eye. In the English translation of Guy de Chauliac's *Cyrurgie*, for example, 'the causes forsothe of sykenesses and of þe sinthomates of the ey3en, after þe more and þe lesse, ben þe firste causes, the causes goynge tofore and þe ioynede causes, as of oþer membres'.[3] For scientific and medical writers, as for Langland, diseased eyes suggest the nature of the whole. In religious contexts,

---

[1] On the self-consciously moralizing impulse of medieval encyclopædists, see Bernard Ribémont, 'On the Definition of an Encyclopædic Genre in the Middle Ages', in *Pre-Modern Encyclopædic Texts: Proceedings of the Second COMERS Congress, Groningen, 1–4 July 1996*, ed. Peter Binkley (Leiden: Brill, 1997), pp. 46-61.

[2] *On the Properties of Things: John Trevisa's Translation of Bartholomæus Anglicus De Proprietatibus Rerum. A Critical Text*, ed. M. C. Seymour et al., 3 vols (Oxford: Clarendon Press, 1975–88), I, p. 178.

[3] *Cyrurgie of Guy de Chauliac*, ed. Margaret S. Ogden, I, EETS, o.s. 265 (Oxford: Oxford University Press, 1971), p. 437.

impeded eyes denote blindness and lack of spiritual light. Eyes and vision are critical to salvation when Christ is 'to knowen us by oure kynde herte and castynge of oure eighen' (XI. 187).[4]

Although the specific ocular conditions of Langland's sins draw upon medical learning, the more general association of poor eye-sight and sin is conventional in the pastoral tradition. Inspired by the preacher's sermon, confession and contrition clear and unlock the penitent's vision. Spoken words – those of preacher and penitent – are transformed into spiritual light. Of course, this is not the case in *Piers Plowman* as the sins remain in a state of obstructed sight. Their conditions are physical, and they do not and cannot transcend the physical body with its limited mode of perception.

Typically, confessions follow sermons. However, the orthodox order is inverted in this passus. James Simpson remarks that 'the sinfulness of the sins is often defined against their abuse of standard ecclesiastical forms which could have saved them from sin'.[5] Pernele's histrionic reaction to Reason's sermon belies any hope of real contrition as does the excessive penance Lechour assigns himself. Coveitise appropriates the confessional language of 'restitucion' for his own ends. Envy's sideways gaze demonstrates 'how the humility and generosity of prayer is grotesquely transformed, and in the movement of Envy's eyes away from the altar how an obsessive concern with earthly standing has eclipsed the spiritual world entirely'.[6] They have not engaged with Reason's sermons, and thus remain in darkness.

Instead, pastoral literature presents effective preaching as a collaboration between listener and speaker. In asking his audience to pray before he begins his St Nicholas Day sermon, a fifteenth-century homilist creates a comparison between himself and his audience to the 'wyse poyetes by olde tyme' who invoke the gods before 'eny grete werk'. The homilist explains that if these pagan poets asked for God's assistance,

> þanne as þus þat gretter werke more nedeful. Ne more trauayle is þere noon þan þe office of prechynge and techyng of godes worde and more ouer syþ

---

[4] William Langland, *The Vision of Piers Plowman*, ed. A. V. C. Schmidt (London: Dent, 1995).

[5] James Simpson, *Piers Plowman: An Introduction to the B-Text* (London: Longman, 1990), pp. 65–66 (and in the second, revised edition (Exeter: University of Exeter Press, 2007), p. 59).

[6] Ibid., p. 66 (2nd rev. edn, p. 60).

þei þat knewe nouȝt þe feyþ creyed to god in worchynge. Moche more scholde we þat schulde be saued by þe feyþ aske helpe of god as he byddyþ.[7]

Although the request for God's help in the success of the sermon is conventional, the parallel drawn between the poet's invocation at the beginning of an epic and the congregation's prayer at the beginning of a sermon invites theorizing upon the pastoral process.[8] Here, as in the case of many other late medieval sermons, the homilist involves his listeners in the unfolding of the sermon:

> I prey ȝou alle at þe bygynnyng of þis schort sarmon preyeþ to god almygti in whom alle witte and wysdom was and his and euere schal be þat he sende me suche grace and schewynge þat hit be hym to worchype and plesyng ȝow to helpe of soule and gostly lyuynge þe deuyl and al his to schame and ouercomyng.[9]

Grace is mediated through the preacher in the delivery of the sermon; however, the audience has a role in its facilitation. Both speaker and listeners are implicated in the experience and ultimate effectiveness of the text—the transformation of the audible words into spiritual light.

It is worth labouring this simple point about collaboration, because modern scholarship has generally disregarded the rhetorical dynamic between speaker and listener. Little criticism concerns the cognitive impact of sermons upon the laity, and the study of exempla has been dominated by a socio-historical focus on their use to contain and to control. With this view, these texts are often simplified into tools for generating fear and submission in which listeners are stripped of any agency.[10] In a recent study of the *Festial*'s exempla, Judy Ann Ford

---

[7] Cambridge, Pembroke College MS 285, fol. 62ᵛ.

[8] By the thirteenth century, it was conventional for the preacher to invite his listeners 'to pray for the good result of the preaching' after introducing the theme. See Phyllis B. Roberts, 'Preaching in / and the Medieval City', in *Medieval Sermons and Society: Cloister, City, University*, ed. Jacqueline Hamesse et al. (Louvain-la-Neuve: Fédération internationale des instituts d'études médiévales, 1998), pp. 151–64 (p. 157).

[9] Cambridge, Pembroke College MS. 285, fol. 63ʳ.

[10] See Jacques Berlioz et al., *Faire croire: modalités de la diffusion et de la réception des message religieux du XIIᵉ au XVᵉ siècle* (Rome: École française de Rome, 1981), Jacques Berlioz et al., *L'aveu, antiquité et moyen-âge: actes de la table ronde* (Paris: École française de Rome, 1986), Jacques Le Goff, 'L'exemplum et la rhétorique de la prédication aux XIIIᵉ et XIVᵉ siècles', in *Retorica e poetica tra i secoli XII e XIV*, ed. Claudio Leonardi and Enrico Mensto (Firenze: La Nuova Italia, 1988), pp. 3–29. Le Goff, for example, argues that late medieval exempla are principally 'au service

reaches a different conclusion. She argues that lay characters are 'distinguished by their exercise of free will' and private interactions with Christ. The identification of 'free will' redresses previous criticism; however, the suggestion that Mirk is 'unique' in this respect seems to overlook similar themes in earlier Latin and English sermon materials.[11] The long and often murky source history of these materials renders such readings difficult to sustain; and despite the efforts of this study and others, any marked theological difference is unconvincing. Exempla often resist historical analysis. Beyond the broadest claims about the shift from the monastic to the pastoral exemplum, they cannot be discussed in terms of linear progression, nor can their variations necessarily be attributed to the agency or manipulation of the writers or compilers.

Despite these critical difficulties, certain things can be said about moral tales and their use in the later Middle Ages. It is clear that they were used with increasing frequency in sermons in keeping with the pastoral reforms of the third and fourth Lateran Councils. Also, exempla were perceived as a form of intellectual mediation for those without clerical education.[12] Finally, within the context of sermons, their ultimate purpose is to encourage the practice of confession.[13] In order for sermon

d'une rhétorique de la peur' (p. 10). Larry Scanlon's *Narrative, Authority and Power: The Medieval Exemplum and the Chaucerian Tradition* (Cambridge: Cambridge University Press, 1994) argues that in the shift from monastic to pastoral models, the sermon exemplum 'will amplify the deference to institutional authority' and 'it will make increasingly abstract the participation it allows in the sacral power of such authority' (p. 65).

[11] Judy Ann Ford's *John Mirk's Festial: Orthodoxy, Lollardy, and the Common People in Fourteenth-Century England* (Cambridge: Brewer, 2006) expands at length upon the private, Christologic experience of the 'Bloody Side' tales. However, this private experience of Christ in exempla pre-dates Mirk. For example, in one very common tale, 'The Bloody Child', lay people see first-hand the impact of their actions upon the body of Christ.

[12] Discussion of the role of exempla in educating the unlearned occurs in medieval *ars praedicadi*, such as Alexander of Ashby's *De modo artificioso predicandi*. He writes that priests must 'dulcem exponere allegoriam et aliquid iocundum enarrare exemplum, ut eruditos delectet allegorice profunditas, et simplices edificet exempli levitas'. Cited in Fritz Kemmler's *Exempla in Context: A Historical and Critical Study of Robert Mannyng of Brunne's 'Handlyng Synne'* (Tübingen: Narr, 1984), p. 71.

[13] On the correlation between sermons and confession, see H. Leith Spencer, *English Preaching in the Late Middle Ages* (Oxford: Clarendon Press, 1993), p. 102 and Alan J. Fletcher, *Preaching, Politics and Poetry in Late-Medieval England* (Dublin: Four Courts Press, 1998), p. 203.

literature to generate confession, the listener must translate a general truth into specific meaning, outer speech to inner dialogue, archetypal exemplum to personal knowledge. In so doing, confessants rely on the mediation of priests and the mediation of sermons and exempla. Likewise, sermons and tales are unsuccessful without listener participation. Pastoral literature thus involves an element of 'active discovery' to borrow a term from modern aesthetic theory. Monroe Beardsley describes 'active discovery' as a 'sense of actively exercising constructive powers of the mind [...] a keyed-up state amounting to exhilaration in seeing connections between percepts and between meanings, a sense [...] of intelligibility'.[14]

Application of the philosophy of aesthetic experience to medieval sermon literature might initially make one bristle for two reasons: firstly, Beardsley was a twentieth-century aesthetician; secondly, pastoral literature is not usually considered 'aesthetic'. However, for Beardsley, 'aesthetic experience' does not presuppose an aesthetic object. Rather, 'exhilaration' and 'intelligibility' aptly describe the transcendental moment of conversion found in confessional tales. These scenes are described in terms of symbolic synæsthesia whereby sensory words converge into spiritual light. Likewise, the 'constructive powers' that enable this transcendence are also suggestive of the intention of confession and the epistemological process that leads there. Listeners to the sermons construct both meaning and their own confessional narratives by engaging with these texts. The following discussion will examine how pastoral literature mediates this process, by examining both self-reflexive passages on preaching found in late medieval sermons and the conversion-by-spoken-word scenes in exempla, using aesthetic experience as an organizing principle.

For an experience to be classified as 'aesthetic' according to Beardsley's philosophy, it must involve four out of five categories. One of these is 'active discovery'. The other four are 'object directedness', 'felt freedom', 'detached affect' and 'wholeness'. While the idea of detached affect is incongruous with *cura animarum*, the other four categories are integrated in the experience of hearing and learning from pastoral texts. Listeners must a) direct themselves to the material in order to achieve b) 'active discovery' by which they experience c) 'wholeness' of their creationary state, which is manifest in the conflation of auditory and visual modes

---

[14] Monroe C. Beardsley, *The Aesthetic Point of View: Selected Essays*, ed. Michael J. Wreen and Donald M. Callen (Ithaca: Cornell University Press, 1982), p. 288.

and d) 'felt freedom', which is the choice of salvation. I shall work through the four relevant categories not in order to prove that pastoral literature conforms perfectly to an anachronistic view of aesthetic experience, but rather as a helpful means of organising the process of reception and understanding in preaching and exempla. 'Aesthetic experience' may offer a corrective paradigm to other work on exempla as it puts emphasis on both perceived and perceiver: speaker and listener collaborate to transform words into light.

Beardsley's last category, 'wholeness', ought to be considered first, because the need for preaching or any form of mediation is directly related to man's lack of wholeness. Pastoral epistemology follows a creationary scheme whereby man is formed, de-formed and re-formed. At Creation, formless matter turns to the Light at the command of the Word whereby the creature is formed.[15] Drawing from a long commentary tradition, late medieval pastoral writers interpret this act as wisdom. Through this wisdom, man was made in God's likeness, a likeness not discerned by corporeal similitude but by reason or 'goostly li3t'. Unlike beasts, men are privileged with reason. As God forms man in his image by reason and with reason, it follows that man deforms himself through lack of reason. This is explained in a fifteenth-century Middle English translation of an Augustinian sermon: 'what man may be more schenschepe to men eþer vnblissful wrecchednes þanne þat is glorie of liknesse of his makere lost. He be drawen to þe *vnschapli* and *vnresounable* likenes of wylde vnresonouable beestes' (emphasis mine).[16] His loss of knowledge is associated with a change not only of intellectual state but metaphorically with a change of physical shape or form.

Although he continues to de-form himself through unreason, man initially experiences the deformation of his created state at the Fall. As one early fifteenth-century homilist writes, 'Adam, instigacione diaboli deformauit ymaginem Dei'.[17] As original sin was believed to be both a verbal and a visual transgression, the resulting deficiencies are realised in hearing and in seeing. In an exegesis of a miracle found in the gospel of Mark, the author of the *Northern Homily Cycle* interprets a 'domb and

---

[15] Augustine, *De Genesi ad litteram*, *CSEL*, 28, I. 1–9.

[16] Cambridge, University Library, MS Ii.6.39, fol. 122ʳ.

[17] 'Adam, through the instigation of the devil, deformed the image of God', quoted and translated in Siegfried Wenzel, *Macaronic Sermons: Bilingualism and Preaching in Late-Medieval England* (Ann Arbor, MI: University of Michigan Press, 1994), p. 294.

defe man' as Adam who 'made vs first fall in blame'. Post-lapsarian Adam and his descendants are metaphorically dismembered as sense perception becomes separated from spiritual understanding: 'To here Goddes word his eres war reft/ Goddes cumandment whan þat he left'.[18]

More frequent than dismemberment is the trope of cloudy or obstructed vision. This obstructed vision often forms the basis for hamartiological schemes, such as that found in Passus V of *Piers Plowman* or in a Quinquagesima sermon in the fifteenth-century sermon collection British Library MS Harl. 331. In this Harley sermon, the general theme of blindness and sin is "dilated" into seven kinds of impediments to vision, which are specific to the seven deadly sins. Pride is likened to smoke, Envy to shadows, Wrath to fiery rock, Avarice to shiny metals, Gluttony to clots of dirt, Lust to stains, Sloth to smoky vapour.[19] In yet another Quinquagesima sermon, the homilist attributes spiritual and physical causes to spiritual and physical blindness. For example, the deleterious effect of old age signifies the state of the soul bemired in sin for a long time; bright light signifies how the good works of others blind the envious; bird droppings signify how lechers are blinded; smoke blinds those affected by worldly honour and pride; dust or powder affects the boastful; and blood the carnally lustful.[20] Elsewhere, certain sins are singled out for association with visual impairments. For example, *The Northern Homily Cycle* warns against 'þe smoke of vanitese' and 'þe pouder of couatise' that 'sum-dele men sese',[21] and *Piers Plowman* 'the smoke and the smolder that smyt in oure eighen,/ That is coveitise and unkyndenesse' (B. XVII. 343-4). *The Orcherd of Syon* also writes of 'þe smoke of pride',[22] likely drawing from biblical references to billowing wind.[23]

Darkened or blinded vision contrasts with the clear showing of the penitent. In order to fulfil this condition, man must clear his spiritual eye from sin to know himself and to know God. As the Middle English

---

[18] *The Northern Homily Cycle: The Expanded Version in MSS Harley 4196 and Cotton Tiberius E vii III*, ed. Saara Nevalinna, 3 vols (Helsinki: Mémoires de la Société Néophilologique, 1972–3, 1984), III, 66.

[19] London, British Library, MS. Harl. 331, fols. 58^r–61^v.

[20] London, Lambeth Palace Library, MS 392, fols. 171^r–174^v.

[21] *The Northern Homily Cycle*, ed., Nevalinna, I, 280.

[22] *The Orcherd of Syon*, ed. Phyllis Hodgson and Gabriel M. Liegey, EETS, o.s. 258 (London: Oxford University Press, 1966), p. 258.

[23] I Timothy 3. 6.

translator of the *Book of Vices and Virtues* explains, 'For riȝt as þe eiȝen of a seke man may not wel loke on þe gret bryȝtnesse, no more may þe vnderstondyng of men as of hemself ne may not loke ne knowe gostliche þinges' until they have been purged of 'tecches of mysbileuynges and of alle oþere filþes'.²⁴ The metaphor has its analogue in patristic commentary. Augustine writes:

> quomodo homo positus in sole caecus, praesens est illi sol, sed ipse soli absens est; sic omnis stultus, omnis iniquus, omnis impius, caecus est corde. Praesens est sapientia, sed cum caeco praesens est, oculis eius absens est: non quia ipse illi absens est, sed quia ipse ab illa absens est. Quid ergo faciat iste? Mundet unde possit uidere Deus. Quomodo si propterea uidere non posset, quia sordidos et saucios oculos haberet, irruente puluere uel pituita uel fumo, diceret illi medicus: Purga de oculo tuo quidquid mali est, ut possis uidere lucem oculorum tuorum. Puluis, pituita, fumus, peccata et iniquitates sunt.²⁵

Despite these visual deformities, sensory learning is redeemed in Christ and the *verbum visibilis* of the sacraments as a means of accessing spiritual understanding. As both human and divine, Christ bridges the gap between the hearing of words and the Word and the seeing of lights and the Light.

Described in sermons for Quinquagesima and other feast days, Christ's miracles restoring sight and hearing position pastoral teaching in the creationary scheme of formation, de-formation and re-formation. Human preaching is specifically redeemed at the moment of Pentecost. In this synæsthetic convergence of word and light – the flaming tongues – human speech achieves a creationary state. In *Word and Light*, David Chidester documents the significance of synæsthesia in the Christian tradition: 'The most important manifestations of the sacred in the

---

[24] *The Book of Vices and Virtues*, ed. W. Nelson Francis, EETS, o.s. 217 (London: Oxford University Press, 1942), p. 222.

[25] Augustine, *In Iohannis euangelium*, *CCSL*, 36 I. 19; trans. John W Rettig, *Tractates on the Gospel of John*, 5 vols (Washington, DC: Catholic University of America Press, 1988): 'Just as when a blind man is placed in the sun, the sun is present to him, but he is absent to the sun, so every slow-witted person, every evil person, every ungodly person is blind in his heart. Wisdom is present; but it is present with a blind man. It is absent to his eyes, not because it is absent to him but because he is absent from it. What then is he to do? Let him cleanse that [his heart] by which God can be seen. Just as if he could not see because he had dirty and sore eyes, with dust or mucus or smoke irritating them, the doctor would say to him, 'Cleanse from your eye whatever foul thing is there that you may be able to see that light of your eyes. Dust, mucus, smoke [these] are sins and iniquities.'

tradition were symbolically structured by the convergence or interpenetration of visual and auditory modes.' Creation, Moses's reception of the Law, the advent of Grace, and the Pentecost mark 'Christian salvation history' as a 'pattern of synæsthetic events'.[26]

The creationary patterns identified in medieval epistemology have not been considered within the context of pastoral literature.[27] Instead, recent study of confessional literature has been focussed upon vision alone, in particular how these texts may be considered in terms of advances in optical theory form the thirteenth century onwards. This method omits the obvious, aural experience of learning, and often draws highly generalised analogies, such as that of vision and understanding, a concept much older in origin.[28] In pastoral literature, successful preaching is associated with synæsthesia, particularly the conflation of audible word and spiritual light. As the incarnate Christ was sent 'sight to preche vnto the blind' – unto the fallen children of Adam – so are earthly priests to do the same with their sermons and exemplary tales.[29] The convergence of senses symbolically restores man to the unity of senses, which was lost at the Fall.[30] As the translator of the *Middle English Mirror* explains of Jesus,

---

[26] David Chidester, *Word and Light: Seeing, Hearing, and Religious Discourse* (Chicago: University of Illinois Press, 1992), p. 21.

[27] In *Word and Light* and 'The Symbolism of Learning in Augustine', *The Harvard Theological Review*, 76 (1983), pp. 73–90, Chidester writes on the symbolic use of word and light in Augustine and other theologians. Marcia Colish's *The Mirror of Language: a Study in the Medieval Theory of Knowledge* (New Haven, MA: Yale University Press, 1968) explores medieval epistemology as relates to language in terms of the Fall and the Incarnation.

[28] The use of vision as an analogy for spiritual understanding is a common trope in patristic, scholastic and pastoral texts. Augustine speaks of the 'heart's eye' in *In Iohannis Euangelium* XV. 19; Bonaventure of the 'eyes of the mind' in *Itinerarium mentis ad deum*, ed. and trans. Philotheus Boehner (St Bonaventure, N.Y.: The Franciscan Institute, 1942), p. 1; and the translator of the *Book of Vices and Virtues* of the 'eien of þe herte' (p. 22).

[29] *The Northern Homily Cycle*, ed. Nevalinna, I, 299.

[30] 'Et uidere et audire simul in verbo est, nec aliud est ibi audire, et aliud uidere sed auditus uisus, et uisus auditus'. Augustine, *In iohannis euangelium*, XVIII. 9; translation by Rettig, *Tractates on the Gospel of John*: 'both to see and to hear are together in the Word; and to see is not one thing, to hear another, but hearing is sight and sight is hearing'.

'his holy werkes wer wordes vnto hem þat vnderstode hem'.[31] Those capable of perceiving synæsthetically—here, those who can 'read' or 'hear' words from visible media—are attributed with a higher form of understanding. As will be treated later, exemplary moments of conversion demonstrate the conflation of auditory and visual modes to signify this higher understanding.

For listeners to see the 'tecches' and 'filþes' that blind them, priests must open their eyes through the medium of spoken words: sermons and exempla. In a state of imperfect and corporeal perception, mediation is necessary, as is the theme of a sermon tale John Mirk borrows from the *Vitas Patrum*. In the story, a leprous king wishes to have 'some maner knowleȝ' of Jesus. As he cannot see him before his own eyes, he commissions a painter to capture his image. 'But when þys paynter lokyt on Crist, hys vysage schon so bryght þat he myght noþyng se of hym'. The painter despairs, but then, 'Cryst toke a cloþe of þys payntur, and wypet his one vysage þerwyth, and þen was þe fowrme of his vysage apertly þeron all oþur'.[32]

Like the painter's cloth, scripture, apostles and sermons serve as mediators of the Light for weakened eyes. Drawing upon the Evangelist's description of himself, commentators explain John as a mediator of this light.[33] Human learning is analogized to a series of reflected lights descending from the ultimate source of light. One homilist explains that 'eche man he liȝtneþ [...] with spiritual wisdom as men þat han þe knowyng of god and þus we may vnderstonde þat all men þat ben liȝtned of hym þei ben liȝtned from whom go forþ al wisdom'.[34]

Preaching is seen as an extension of this light. Citing Grosseteste, a fifteenth-century homilist discusses the Church as a great light: 'iam diffundit suum lumen in terram humane per prechinge and techinge'.[35] Likewise, the *Middle English Mirror* interprets the healing of a blind man

---

[31] *The Middle English Mirror: Sermons from Advent to Sexagesima*, ed. Thomas G. Duncan and Margaret Connolly, Middle English Texts, 35 (Heidelberg: C. Winter, 2003), p. 85.

[32] John Mirk, *Mirk's Festial: a Collection of Homilies*, ed. Theodor Erbe, EETS, o.s. 96 (London: Keegan Paul, Trench & Trübner, 1936), p. 264.

[33] John 5. 35.

[34] London, British Library, Harl. MS 2276, fol. 18ʳ.

[35] '[Ecclesia] now sheds its light on the earth of the human soul by preaching and teaching'. 'De celo querebant' quoted and translated in Wenzel, *Macaronic Sermons*, p. 279.

as the preaching of a priest: 'Jesus dide brynge þe blynde man to hym. & þe prest schal laden [þe synful] þoruȝ prechynge [...] & þoruȝ penaunce, vnto Iesus þat schal aliȝtten hym'.[36] Just as the blind man showed Christ that he was blind, so the listeners are asked to reveal their sins so they may be illuminated. Although accomplished through temporal and sensory means, effective preaching leads to an eternal and incorporeal light.

In the pastoral tradition, penitents are able to recognise their own sins to achieve this clear vision. Given that the 'tecches' and 'filþes' – the sins themselves – are visual impediments, they need external help either through their confessors or pastoral literature, to see how sins impede this spiritual vision. In the *Speculum Guy de Warwycke*, for example:

> And, whan mannes soule, ful iwis,
> Þurw dedli sinne ifiled is,
> His knowelaching is al gon;
> For wit ne *siht* haþ [he] non.[37]

The aim of experiencing of pastoral literature is thus a kind of wholeness, both in terms of eventual unity of God, and a unity of perception.

However, listeners are not passive receivers, and this leads to Beardsley's criterion of 'object directedness'. In assuming that man can learn spiritual matters from sensible words, pastoral texts are consistent with medieval traditions of hierarchical perception whereby man progresses from corporeal to spiritual to intellectual vision or from the sensual to the self to the eternal.[38] Within these schemes, however, the presence of will is necessary in the ascent from sensible to intellectual perception.[39] Although vision is considered to be the most privileged sense for its element of wilful selectivity—one chooses what one seeks in a frame whereas one cannot choose what one hears—pastoral writers extend a similar argument in regards to listening.

---

[36] *The Middle English Mirror: an Edition Based on Bodleian Library, MS Holkham misc. 40*, ed. Kathleen Marie Blumreich (Tempe, AZ: Arizona Center for Medieval and Renaissance Studies, in association with Brepols, 2002), p. 108. (This version is consulted as the Duncan and Connolly edition is not complete.)

[37] *Speculum Guy de Warewyke*, ed. Georgiana Lea Morrill, EETS, e.s. 75 (London: Keegan Paul, Trench & Trübner, 1898), p. 33 (emphasis mine).

[38] See Augustine, *De Genesi ad litteram* XII. 6 and Bonaventure, *Itinerarium mentis in deum*, p. 46.

[39] See Margaret Ruth Miles, 'Vision: The Eye of the Body and the Eye of the Mind in Saint Augustine's *De trinitate* and *Confessions*', *The Journal of Religion*, 63 (1983), 125–42.

Listeners must choose actively to accept the mediation of sermons and tales in order for them to have effect. *The Middle English Mirror* speaks of deafness in these terms. While physically capable of hearing, some people refuse to listen and to process the words spoken: 'he is def þat hereþ þe gode and doþ no þyng þeraftur'. Just as Augustine writes in terms of selectivity of vision, pastoral writers discuss selectivity in hearing: 'hii ne wyl noþinge heren but þat hem lykeþ'.[40] Similarly, other passages suggest the necessity of discretion based on the content of the priest's sermon: 'Noman ne schal noman forsaken þat spekeþ Goddes worde, ac wiþ gode wille heren him & al her wordes iuggen wel wiþin her hertes, & leten þe iuel & taken þe gode'.[41] This theme is not unique to *The Mirror*. *The Northern Homily Cycle*, for example, urges the laity to judge the speech of prelates and priests in order to determine whether to listen further or to act upon their advice:

> To prelatis owe we buxumnes
> ȝif his bidding falle to goodnes
> But if he bidde vs don ille
> We owe not to don his wille.[42]

This advice is followed by the relevant exemplum of a hermit who improperly counsels a novice.

Not only must listeners physically hear, but they must actively engage with the sermons in order to achieve self-knowledge and salvation. Those who simply let the words wash over them may as well sit at home. A fifteenth-century sermon distinguishes between three kinds of ineffectual minds: the first whose thoughts are burdened with 'vain thoughts'; the second who listen with joy but then quickly forget when the sermon is over; and the third who are too distracted with worldly cares for the sermon to have effect.[43] Pre-occupation with other matters impedes the synæsthetic transfer of sensual words into spiritual light. The compiler of the *Northern Homily Cycle*, for example, writes of those who cannot 'hear' the sermon, because their inner vision is blocked by thoughts of worldliness. The 'riche man þat sarmon heres' cannot benefit from it

---

[40] *The Middle English Mirror*, ed. Duncan and Connolly, p. 43.
[41] Ibid., p. 19.
[42] *The Northern Homily Cycle*, ed. Anne B. Thompson (Kalamazoo, MI: Medieval Institute Publications, 2008), p. 111. (This edition is cited as the sermon does not occur in the expanded text previously cited.)
[43] Oxford, Bodleian Library, University College MS 28, fol. 62ʳ.

because 'his hert al couert es/ With couetise and besines'.⁴⁴ Many men have spiritually impeded eyes as noted above, but they must listen and engage with pastoral teaching in order to clear their vision.

How sermons and exempla mediate self-knowledge once they have the object directedness of the audience leads to the criterion of active discovery. By actively engaging with sermons and exempla, congregants in conversion exempla are re-formed. As man's de-formation is represented by blindness and deafness – the loss of 'right spekeing' and 'hereing of ere' in *The Northern Homily Cycle* – his reformation is represented as not only a restoration but a convergence of hearing and seeing. The words that the 'prechore bringes fro Goddes horde' unlock the spiritual vision of the penitent, allowing him to 'schew his sins in schrift'.⁴⁵

At another level of synæsthetic learning, confessional narratives themselves are a means of expressing invisible reformation through 'visible' means. For example, in the version of the 'Incestuous Daughter' found in *Jacob's Well*, words are made visible to signify her conversion and return to God. The daughter is so moved by a sermon 'in þo woordys of þe frere, here herte braste for sorwe of here synnes, & dyed'. She interprets the sensory words in personal terms: recognition of her sins. Although she does not have a chance to physically confess, her contrition is sufficient for her to return to God. Her inner conversion is made known to onlookers in a visible sign. When she is buried a 'fayr tre' sprouts from her grave 'wretyn aboute in euery leef, wyth letters of gold' displaying words of scripture in Latin.⁴⁶

Spiritual illumination is not the exclusive property to those of profound intellectual capacity or clerical learning. A similar Marian exemplum concerns a knight who becomes a monk out of his devotion to the Virgin. Despite the brothers' efforts to teach him Latin and 'þe buke', he can only successfully learn two words, 'Ave Maria'. When he dies, a lily sprouts from his mouth 'an on evur-ilk a lefe þeroff was wreten, 'Ave Marie', with gold lettres'. Despite the monks' erudition, it is through these visible words pertaining to an unlettered man that 'oure ladie wold þat lat þaim hafe knowledge'.⁴⁷ In the *Speculum sacerdotale* version, which

---

⁴⁴ *Northern Homily Cycle*, ed., Nevalinna, II, 21.
⁴⁵ *Northern Homily Cycle*, ed., Nevalinna, III, 68.
⁴⁶ *Jacob's Well*, ed. Arthur Brandeis, EETS, o.s. 115 (London: Oxford University Press, 1900), p. 173.
⁴⁷ *An Alphabet of Tales*, ed. Mary Macleod Banks, EETS, o.s. 126 and 127 (London: Keegan Paul, Trench & Trübner, 1905–5), p. 53.

follows closely that of the *Alphabet of Tales*, the tale concludes, 'and then they vnderstode with what deuocion he saide the forsaide two wordys, therfore God hadde liȝtyned hym with siche worschipe and honowre'.[48] While the gold letters literally illuminate his grave, the knight is also spiritually illuminated in the knowledge of God's grace. In other tales, auricular confession alters external visions and forms. All of these tales present the mediator in an important role. While, as has been suggested by critics, many exempla represent a topical urgency in establishing priestly authority in the administration of sacraments, priests also serve as foils for the listeners and readers of the tales.[49] In the reading of faces and bodies, priests mediate the hermeneutic process that all penitents must undertake in order to re-create their own narratives.

In these tales, sins are 'read' on bodies, and spoken words make visual impact. A priest in *Handlyng Synne* is able to 'knowe þe god from þe yl' by reading his parishioners' faces. While some of their faces are 'bryȝt', many others reflect their sinfulness:

> And some were as rede as blode,
> Staryng ryȝt as þey had be wode;
> And sum were swolle, þe vyseges stout,
> As þoȝ here yȝen shulde burble out;
> And sum gnapped here fete & handes,
> As dogges doun þat gnawe here bandes;
> And sum hadde vysages of meslrye[50]

God allows the priest to interpret these visible signs as texts revealing specific sins.

Penitents actively reform these texts through confession. In *Speculum sacerdotale*, for example, a hermit sees a man 'blak and clowdy' emerge from church with 'a white face and cliere in alle parties'. The hermit immediately interprets his new physical form. In asking the man to recount his past life and how he was 'conuertyde', he recognises the existence of two narratives symbolised in his two forms. The man recounts his tale of hearing the scripture read in a sermon. The sermon

---

[48] *Speculum sacerdotale*, ed. Edward H. Weatherly, EETS, o.s. 200 (London: Oxford University Press, 1936), p. 48.

[49] On the preoccupation with church ritual and authority in exempla, see Scanlon, *Narrative, Authority and Power*, p. 70.

[50] Robert Mannyng, *Robert of Brunne's 'Handlyng Synne'*, ed. Frederick J. Furnivall, EETS, o.s. 119, 123 (London: Oxford University Press, 1901), p. 318.

from Isaiah serves as a mirror. The mediated word allows him to *see* himself spiritually. He quotes some of the verse, explaining, 'and with that I *seeying* that I was a gret synful wrecche and a gret fornicatoure began to be contrite and to be sorowefull in herte for my synne at the movynge of this swete sermone' (emphasis mine).[51] Through this mediation, he is able to reform himself and construct a confessional narrative for the benefit of others.

Whereas the 'Incestuous Daughter' does not have the opportunity to confess, her listening process is similar. The sermon 'into her body liʒt'. Although 'liʒt' is a variant of the verb 'alighten', it effectively resembles the noun. The entrance of the words into the body where they are transformed into a new substance is reminiscent of medieval descriptions of the Annunciation, that most powerful example of synæsthetic convergence and creation.[52] The words having penetrated her, she asks the bishop 'what may þis be?/ Alle day þou hast spoken of me'.[53] The sensory words of the sermon are actively constructed in terms of her own experience, resulting in acute remorse and desire for absolution. In the version of the tale found in Bodley 649, the priest sees the woman led by devils and chains while preaching and 'vertebat materiam sui sermonis in ista mouencia verba misericordie'.[54] However, the words that he utters do not specifically relate to her situation, but rather describe the suffering of the passion and the mercy this releases into the world, and exhort wretched sinners to confess. That these words move her to contrition suggests how penitents may apply the general words of sermons to their specific circumstances. In the woman's case, she cannot see the chains that bind her – materially represented to the priest as chains, but symbolically, of course, by her despair – until the sermon unlocks her vision. The transformation wrought by this process is revealed to the parishioners by an angel: she 'schynes in heon as son brith'.

Through 'active discovery' in listening to pastoral texts, listeners experience wholeness of perception and what might be called 'felt

---

[51] *Speculum sacerdotale*, p. 59.
[52] *MED* 'alighten', v. (1), 3 (a).
[53] 'A Critical Edition of Cambridge University MS Ff.5.48' ed. J. Y. Downing (unpub. Ph.D. diss., University of Washington, 1969), p. 113.
[54] '[He] turned the matter of his sermon in to these moving words of mercy', ed. and trans. Patrick Horner, *A Macaronic Sermon Collection from Late Medieval England: Oxford, MS. Bodley 649* (Toronto: Pontifical Institute of Mediaeval Studies, 2006), pp. 148–49.

freedom'. For Beardsley, this criterion describes 'a sense of release from the dominance of some antecedent concerns about past and future'.[55]

In the conventional language of confessional exempla, penitents are quite literally freed from the chains and bonds of the Devil. As the spiritual eyes of the listeners are cleared and illuminated, the spiritual eye of the devil is put out or darkened. Many confessional tales describe devils that can no longer see sinners once they have performed confession. The penitent literally becomes 'vnkouth' or unknown to the Devil through his self-knowledge. As the *Northern Homily Cycle* explains, 'schrift of mouth/ Mase man saul to þe fend vnkouth'.[56] An exemplum found in Mannyng's *Handlyng Synne* illustrates the seventh grace of shrift, which 'the deuyl blyndeþ'. As in other tales, the sinner previously led by the Devil in chains becomes invisible after auricular confession. The prologue to the tale explains this within the post-lapsarian scheme of human knowledge and perception. Confession puts out the 'yȝe gostly' of the Devil just as he 'refte vs alle gostely syȝt' at the Fall.[57] In addition to blinding men, the devil also binds man's 'tung and eres', limiting their ascent to the Word and Light. In their freedom from both the Devil's chains and limited perception, penitents are able to ascend intellectually and gravitationally.

Exemplary conversion tales, in combination with homiletic explanations of active and passive listening, outline a pattern of re-form whereby listeners engage with texts cognitively and constructively. In this design, pastoral literature assigns a dignity to the individual penitent that is often overlooked in modern scholarship. These texts do not simply rely on fear of other-worldly punishments to control the masses, but rather stress the capacity of penitents to learn and ascend through the intellect. The verbal associations of Creation – word, light and learning – resonated with symbolic understanding for a medieval audience. By placing the human experience of learning within this creationary scheme, listeners are able to participate in one of the most important events of human history.

---

[55] Beardsley, *The Aesthetic Point of View*, p. 288.
[56] *Northern Homily Cycle*, I, 146.
[57] Mannyng, *Handlyng Synne*, pp. 379–82.

Sights for Sore Eyes:
Vision and Health in Medieval England

Joy Hawkins

In his celebrated *Chirurgia*, Henri de Mondeville (d. *c*.1320), surgeon to King Philip the Fair of France, noted that:

> The eye being the noblest and most frail of the external organs, a single grain of corrosive matter is more harmful to it than a hundred would be to feet or jaws; thus we must operate on the eye and such like organs much more carefully.[1]

Henri's work was quickly translated into a number of vernacular languages, and became popular amongst English medical practitioners who recognised that damage to the eyes caused more pain and discomfort than other physical injuries. It is not surprising, therefore, that the number of remedies focusing on ocular complaints in medical books and herbals was particularly high. In one single fifteenth-century manuscript entitled 'A Litel Boke of Medicyns', now bound in British Library MS Sloane 405, there are just over 250 recipes for bodily ills, of which 32 concentrate on ailments of the eyes, a number only equalled by cures for digestive problems.[2] Together, these two types of disorder account for over a quarter of the entries in the 'Litel Boke', suggesting that the health of the eyes, alongside that of the digestion, caused the greatest concern.[3] The eye was, of course, an easily accessible part of the body, which could often be treated effectively with comparatively simple preparations, which

---

[1] Marie-Christine Pouchelle, *The Body and Surgery in the Middle Ages*, trans. Rosemary Morris (New Brunswick, NJ: Rutgers University Press, 1990), p. 119.

[2] BL, MS Sloane 405, fols 126ʳ–154ᵛ. For 'De oculis', see fols 129ʳ–131ᵛ.

[3] Ailments of the eyes and the digestive system were interconnected. As the physician Gilbertus Anglicus (*fl.* 1240) noted, many diseases of the eyes had their origins in the poor digestion of food: 'ache of þe yȝen comeþ otherwhiles from þe stomake' (*Healing and Society in Medieval England: A Middle English Translation of the Pharmaceutical Writings of Gilbertus Anglicus*, ed. Faye M. Getz (Madison, WI: University of Wisconsin Press, 1991), p. 33).

helps to explain the profusion of remedies in these pages.[4] Loss of sight was profoundly debilitating in both physiological and economic terms and thus to be avoided at all costs. Furthermore, as we will see, visual impediments caused additional anxiety amongst medically-informed patients because of the threat they posed to the well-being of the entire body. It must be noted that blindness could be equally regarded as a gift from God, conferring inner wisdom and providing the opportunity for personal redemption.[5] Good Christians who accepted their infirmity with humility hoped that they would be spared a far worse ordeal in purgatory.[6]

This article will explore the effect that sight was thought to have on the whole physiological system. As we shall see, looking upon beautiful objects, such as colourful flowers and green grass, fortified the spirits and kept them healthy, while disgusting or disturbing images could corrupt the body, and, *in extremis*, threaten life itself. Through an examination of two of the commonest causes of ophthalmic complaints – old age and performing close, detailed work in poor light – we will consider how far the introduction of spectacles helped long-sighted craftsmen and scribes.[7] None the less, it will be suggested that, whatever the physical cause, blindness had grievous consequences for a person's spiritual well-being, confining the individual to a private world of darkness, and preventing him or her from witnessing the elevation of the Host or benefiting from the therapeutic influence of religious iconography. The eyes were thought to present a physical manifestation of the state of the soul, which meant that any inner spiritual deformity would reveal itself in them. Red, sore, cloudy or weeping eyes, for instance, could signify a sinful soul that needed purgation. Having investigated such beliefs, this article will also seek to understand the preventative measures that medieval men and

[4] Tony Hunt, *Popular Medicine in Thirteenth Century England: Introduction and Texts* (Cambridge: Brewer, 1990), pp. 264–67.

[5] Joy Hawkins, 'Seeing the Light? Blindness and Sanctity in Later Medieval England', in *Saints and Sanctity*, ed. Peter Clarke and Tony Claydon, Studies in Church History, 47 (Woodbridge: The Ecclesiastical History Society by the Boydell Press, 2011), pp. 148-58.

[6] Carole Rawcliffe, *Leprosy in Medieval England* (Woodbridge: Boydell, 2006), pp. 55–6, 141.

[7] For a full discussion of the other possible causes of blindness and eye complaints, see Joy Hawkins, 'The Blind in Later Medieval England: Medical, Social and Religious Responses' (unpublished PhD thesis, University of East Anglia, 2011), pp. 53-88.

women took to protect the most precious of their sensory organs. Before we can consider the relationship between the condition of the eyes and the health of the body, however, we first need to review briefly medieval concepts of ocular physiology.

Initially, there was some disagreement regarding how the process of sight actually worked. As medieval scholars and physicians studied the newly available translations of the works of the Ancients, they became divided between the Galenic and Aristotelian theories of vision. Aristotle (384-322 BC) 'assigned the observer a passive role in vision', believing that objects sent out rays through the air to the eye, which absorbed them.[8] A chain of images or impressions was created, whereby '[the] shape of things give shape to the air that intervenes between themselves and the eyes'.[9] On the other hand, the eminent Greek physician, Galen (c.129-200/216), argued that the eye was a far more active agent, sending out an intangible substance, called the visual spirit, to the object so that it could return with an exact copy which was then transported through two hollow tubes or optic nerves to the front ventricle of the brain.[10] The visual spirit 'thus [became] an extension of the optic nerve and an instrument of the soul'.[11] This spirit was eloquently described as a 'fiery force', which passed 'first through the eyes, and then to the objects to be seen, and by returning to its point of origin [brought] back to the mind [...] the shape impressed upon it as though by a potter'.[12] Put simply, the visual spirit appeared to carry all such imprints from the crystalline lens of the eye to the nerve receptors at the front of the brain, where the act of vision was completed. It followed logically that any image entering the

---

[8] David C. Lindberg, 'The Science of Optics', in *Science in the Middle Ages*, ed. idem (Chicago: University of Chicago Press, 1978), pp. 338–68 (pp. 340–41).

[9] Adelard of Bath, 'Natural Questions', trans. Hermann Gollancz, in *A Sourcebook of Medieval Science*, ed. Edward Grant (Cambridge, MA: Harvard University Press, 1974), p. 377.

[10] E. Ruth Harvey, *The Inward Wits: Psychological Theory in the Middle Ages and Renaissance* (London: Warburg Institute, University of London, 1975), pp. 42–43. *On the Properties of Things: John Trevisa's Translation of Bartholomaeus Anglicus' De proprietabus rerum*, ed. Michael C. Seymour et al., 3 vols (Oxford: Clarendon Press, 1975–1988), I, 98.

[11] Lindberg, 'The Science of Optics', p. 341.

[12] Adelard of Bath, 'Natural Questions', in *A Sourcebook of Medieval Science*, p. 377; *On the Properties of Things*, I, 108.

body from the outside world could change its entire equilibrium, either for the better or the worse.[13]

These theories about the physiology of the eye were transmitted to the West from the tenth century onwards through the work of respected Muslim authorities such as Haly Abbas (d. 994/5) and Avicenna (d. 1037). As Christian theologians and medical practitioners were exposed to the newly available texts emerging from the recently established schools at Toledo and Salerno, they adapted them into their own work.[14] Following the translation of a growing corpus of material from Arabic and Syrian into Latin, the ideas of Galen and Aristotle, alongside those of Plato (428–347 BC) and Euclid (c.325–265 BC), were rationalised and amalgamated to achieve a more consistent theory of optics.[15] We can, therefore, understand why scholars, such as Robert Grosseteste (1170–1253) and Roger Bacon (1214–1292), reached the paradoxical conclusion that sight could be both passive and active.[16] Men trained in the universities that were springing up across Europe understood optics within the theoretical context of geometry, which formed a significant part of the arts syllabus.[17] Consequently, their approach was mathematical rather than physiological, which is clearly demonstrated in the detailed diagrams that accompanied their work. Not only did they analyse and describe the different parts of the eye with technical precision, they also concentrated on identifying the angles at which light rays entered the eye in various circumstances.[18] Although these new ideas

---

[13] Harvey, *The Inward Wits*, pp. 17–18.

[14] David C. Lindberg, *The Beginnings of Western Science: The European Scientific Tradition in Philosophical, Religious, and Institutional Context, 600 BC – AD 1450* (Chicago: University of Chicago Press, 1992), p. 170.

[15] Lindberg, 'The Science of Optics', pp. 349–51.

[16] Suzannah Biernoff, *Sight and Embodiment in the Middle Ages* (Basingstoke: Palgrave, 2005), p. 71.

[17] Faye M. Getz, 'The Faculty of Medicine before 1500', in *The History of the University of Oxford, Volume II: Late Medieval Oxford*, ed. Jeremy I. Catto and Ralph Evans (Oxford: Oxford University Press, 1992), pp. 373–413. For slightly later evidence of the study of geometry at Cambridge, see Damian R. Leader, *A History of the University of Cambridge, Volume I: The University to 1546* (Cambridge: Cambridge University Press, 1988), pp. 144–45.

[18] Peter Murray Jones, *Medieval Medical Miniatures* (London: British Library in association with the Wellcome Institute for the History of Medicine, 1984), pp. 49–50. Richard W. Southern, *Robert Grosseteste: The Growth of an English Mind in Medieval Europe* (Oxford: Clarendon Press, 1986), p. 158. Roger Bacon, *Opus*

would have been accessible to the better-educated clergy, to elite physicians and to the users of monastic libraries, other medical practitioners and laymen would not initially have been exposed to such complex and potentially challenging concepts. They did, however, become increasingly aware of the dramatic effect that images might have upon health, and would naturally have been concerned at a more pragmatic level about the spiritual and physical implications of lost sight.

Since knowledge of the circulation of the blood lay far in the future, medieval physicians followed Galen in comparing the body's complex network of veins and arteries to an irrigation system, whereby essential matter was conveyed to the extremities through the medium of three interconnected fluids or spirits. The venous blood, or natural spirit, transported a mixture of the four humours (phlegm, black and yellow bile and blood), which fed and nurtured the whole body. The arterial blood, or vital spirit, was formed from a mixture of inhaled air and venous blood which appeared to have been filtered through the porous septum in the heart. Also known as *pneuma*, the hot, frothy substance produced during this process was carried along the arteries bearing heat. Some of it travelled upwards towards the brain, where it was filtered again through the *rete mirabile*, which Galen (erroneously) located at the base of the skull. When this superior, refined blood mingled with air from the nostrils, the animal spirits were generated in the brain. They were the agents which made movement and thought possible and were immediately responsive to external stimuli.[19] From these developed the aforementioned visual spirit, which carried the images it had gathered from the outside world along the hollow optic nerves to the brain.[20]

Both Aristotle and Galen maintained that when an object was illuminated it radiated a likeness of itself into the air, like a snake shedding its skin, which was known in Latin as *simulacrum*.[21] Each *simulacrum* was identical in every way to the original and possessed the same humoral make-up; consequently, when a copy of the object viewed was absorbed into the body its innate qualities were also assimilated. All

---

majus, trans. Robert Belle Burke, 2 vols (New York: Russell & Russell, 1962), II, 432, 442, 448, 501–05.

[19] Harvey, *The Inward Wits*, pp. 4–25; Rawcliffe, *Leprosy*, pp. 67–71.

[20] Harvey, *The Inward Wits*, pp. 18, 43. Nancy Siraisi, *Medieval and Early Renaissance Medicine: An Introduction to Knowledge and Practice* (Chicago: University of Chicago Press, 1990), p. 108.

[21] Lindberg, 'The Science of Optics', p. 340.

natural substances, including animals, plants, metals and minerals, were believed to have a unique complexion formed from a combination of heat, cold, moisture and aridity.[22] Everything one ate, drank, viewed or smelt had an impact on the physiological system. For instance, gazing upon the colour red, eating leeks or inhaling the aroma of fennel had a heating and drying effect on the body.[23] It is worth noting that the processes of vision and olfaction appeared to be very similar. Not only did objects release a visual *simulacrum* of themselves into the air, they also emitted an odour, described by the thirteenth-century encyclopaedist Bartholomaeus Anglicus as a smoky substance existing somewhere between air and water.[24] Odours, like *simulacra*, were thought to possess the qualities and degrees 'of the substance from which they emanated – they could be hot or cold, wet or dry'.[25] These qualities were likewise assimilated into the body and transported to the brain, where they interacted with the animal spirits. As was the case with conflicting theories about vision, during the Middle Ages there was some debate regarding whether the nose, the nasal nodules, the brain or the odour itself constituted the active agent.[26]

So far as the process of vision was concerned, whether the rays themselves transported the 'visible qualities' of the object 'through the intervening air',[27] or whether the eye itself sent out an active force to interact with the *simulacra*,[28] was a question of little practical significance

[22] Siraisi, *Medieval and Early Renaissance Medicine*, pp. 101–04. These qualities were further classified by Galenic theory according to a sophisticated gradated system of four levels of heat and moisture. The first grade was mild whilst the fourth was extreme, and even potentially lethal. This system of ordering had been known in medieval Europe since the appearance of Constantine the African's translation of the *Liber graduum*: Michael McVaugh, 'An early discussion of medicinal degrees at Montpellier by Henry of Winchester', *Bulletin of the History of Medicine*, 49 (1975), 51–71 (pp. 60–61).

[23] 'The Gouernaunce of Prynces or Pryvete of Pryveteis', in *Three Prose Versions of the Secreta Secretorum*, ed. Robert Steele, EETS, e.s. 74 (London: Keegan Paul, Trench & Trübner, 1898), p. 244. *The Medieval Health Handbook: Tacuinum Sanitatis*, ed. L. Cogliati Arano (New York: Braziller, 1976), plates XIII, XXVIII.

[24] *On the Properties of Things*, II, 864.

[25] R. Palmer, 'In Bad Odour: Smell and its Significance in Medicine from Antiquity to the Seventeenth Century', in *Medicine and the Five Senses*, ed. W. F. Bynum and Roy Porter (Cambridge: Cambridge University Press, 1993), pp. 61–68 (p. 63).

[26] Ibid., pp. 62–64. Harvey, *The Inward Wits*, p. 6.

[27] Lindberg, 'Science of Optics', pp. 340–41.

[28] Harvey, *The Inward Wits*, p. 42.

in terms of the physiological impact upon the viewer. In both instances the *simulacrum* was ultimately absorbed into the eye and conveyed directly to the brain, which was the home of the animal spirits. If the image in question was disgusting, such as a rotting carcass, a noisome dung heap or even an ugly, diseased person, it would have a detrimental effect, potentially destabilising the entire system. The animal and vital spirits were sensitive creatures which could easily become disturbed, rushing in and out of the principal organs (respectively the brain and heart where they dwelt) and creating havoc everywhere. If the animal spirits, which carried sensory information down the spinal cord, moved too rapidly, reacting in shock to what had been seen, they could induce fainting, paralysis or collapse. Moreover, because each *simulacrum* replicated the object's qualities, an excessively strong phlegmatic or melancholic image could harm the body from within. If the vital spirit experienced extreme fear or nausea and suddenly retreated into the heart, it could diminish the body's innate heat, or life spark, leading, at best, to a deep sense of dread, and in extreme cases actually causing death. Conversely, flight to the extremities could be equally injurious to the natural thermostat.[29]

On the other hand, although unpleasant sights could prove fatal, pleasing ones had the opposite effect. The heartening prospect of a garden laid out with lawns, flowers and statues, or even a handsome person wearing fine clothes, fortified the entire system by strengthening the spirits.[30] By the same token, exposure to strong, vibrant colours, especially in the form of precious stones, which were thought to emit healing rays, appeared to energise the spirits and prevent disease. Many cures for blindness harnessed the occult powers of such minerals. Bartholomaeus Anglicus deemed the subject of gems sufficiently important to devote a whole book to them in his celebrated encyclopaedia, presenting evidence from a variety of sources, including Isidore of Seville (*c.*560–636) and the Greek pharmacologist, Dioscorides (*c.*40–90), famous for his *De materia medica*.[31] Specifically, Bartholomaeus recommended 'prassius [...] a stoon

---

[29] Harvey, *The Inward Wits*, p. 17.
[30] Mahmoud A. Manzalaoui, *Secretum secretorum: Nine English Versions*, EETS, e.s. 276 (Oxford: Oxford University Press, 1977), p. 5.
[31] See, for example, *On the Properties of Things*, II, pp. 840–1, 852–5, 858–73; Joan Evans, *Magical Jewels of the Middle Ages and the Renaissance* (Oxford: Clarendon Press, 1922; repr. New York: Dover, 1976), pp. 90–91.

grene as leeke' to rectify 'feble sight'.[32] He also praised the therapeutic effects of sapphire, which was thought to possess particular properties that protected the sight, largely by analogy because it was the 'chief of precious stones', just as the eye was considered to be the first of the senses, as well as one of the noblest organs in the body.[33] Peter of Spain (d. 1277), who later became Pope John XXI, also claimed that these gems healed diseases of the eye.[34] One stone in particular, referred to as 'tutty' or 'totye' in medieval manuscripts, now known as zinc oxide, was frequently prescribed for sore eyes.[35] For example, one entry in a fifteenth-century 'Book of Simples' instructs the reader to take the stone 'þat men callis totye' and mix it with fresh water to make an ointment.[36] Some precious stones had to be powered to release their medicinal properties; it was believed that jasper, when pulverised, would cleanse 'þe yhen of hore and of filthe, and scharpeþ and comforteþ þe sight'.[37] On other occasions, however, the close proximity of precious stones was deemed sufficient, simply by virtue of their remarkable aura. Consequently, gems and jewels were often made into amulets to shield the wearer from danger. In Chaucer's version of *The Romaunt of the Rose*, 'a stoon full precious, / That was so fyn and vertuous / That hol a man it koude make' was attached to the hero's belt buckle, safeguarding him against palsy, toothache and blindness.[38]

A fifteenth-century version of the popular *Regimen sanitatis* (guide to healthy living) underscored the importance of being surrounded by suitably uplifting visual stimuli:

[32] *On the Properties of Things*, II, 864; R. E. Latham, *Revised Medieval Latin Word-List from British and Irish Sources with Supplement* (London: Published for the British Academy by Oxford University Press, 1965).

[33] Pouchelle, *Body and Surgery*, pp. 119–21. *On the Properties of Things*, II, pp. 869–70.

[34] Evans, *Magical Jewels*, p. 113. Petrus Hyspanus (Pope John XXI), *The Treasury of Health* (London, c.1550), cap. vii.

[35] Getz, *Healing and Society*, p. 358; W. J. Daly and R. D. Yee, 'The Eye Book of Master Peter of Spain: A glimpse of diagnosis and treatment of eye disease in the Middle Ages', *Documenta Ophthalmologica*, 103 (2001), 119–53 (pp. 130, 134).

[36] BL, MS Lansdowne 680, fols 39ᵛ–40ʳ.

[37] *On the Properties of Things*, II, p. 853.

[38] Geoffrey Chaucer, 'The Romaunt of the Rose', in *The Riverside Chaucer*, ed. Larry D. Benson (Oxford: Oxford University Press, 1988), ll. 1085–1102. For a general discussion of amulets worn for medical purposes, see Valerie I. J. Flint, *The Rise of Magic in Early Medieval Europe* (Oxford: Clarendon Press, 1991), pp. 301–4, 310.

> Se that thi clothis be precious and ri3t feire to the eye, for beauté and preciousenes of þe clothis li3tenith and gladdith the spiritte of man, which gladness of spiritte is cause of a continuaunce in helth like as heuynes of spiritte and sorow inducith siknes.³⁹

Many such *regimina* began by emphasising that a strong constitution was a precious commodity, as once lost it would be difficult, if not impossible, to regain.⁴⁰ Copious advice was provided to guide the reader, who was urged to engage in such therapeutic activities as laughter, drinking good wine (in moderation), listening to music or walking in a garden. The thirteenth-century Dominican bishop and scholar, Albertus Magnus, who in later life became a key figure in the development of medieval optical theory,⁴¹ regarded the pleasure garden as an ideal place for maintaining physical as well as mental equilibrium:

> Nothing restores the sight so much as fine short grass [... therefore] a great diversity of medicinal and scented herbs [should be planted] not only to delight the sense of smell by their perfume but to refresh the sight with the variety of their flowers, and to cause admiration at their many forms in those who gaze upon them.⁴²

It was the colour of the grass which constituted its principal virtue. Green was the ideal, median colour, 'ygendred bytwene rede and blak'.⁴³ It appeared, therefore, 'most liking to þe sight' and seemed to comfort 'þe visible spirit', restoring its natural equilibrium.⁴⁴ According to the *Cause et cure* of the influential Benedictine abbess and polymath, Hildegard of Bingen (1098–1179),

> if, due to old age, or some infirmity, blood and water have significantly decreased in a person's eyes, he should go to a green meadow and look at it long enough until the eyes become moist as from weeping, because the

---

³⁹ Manzalaoui, *Secretum secretorum*, p. 4.
⁴⁰ Ibid., pp. 144, 221.
⁴¹ Lindberg, 'Science of Optics', pp. 349–50.
⁴² Albertus Magnus, *De vegetabilibus libri VII*, ed. Ernest Meyer (Berlin: G. Reimer, 1867), cap. XIV, pp. 636–37.
⁴³ *On the Properties of Things*, II, p. 1290. According to the philosopher and theologian William of Auvergne (d. 1249), green 'lies half way between white, which dilates the eye, and black, which makes it contract' (quoted in Umberto Eco, *Art and Beauty in the Middle Ages*, trans. Hugh Bredin (New Haven & London: Yale University Press, 1986), p. 46).
⁴⁴ *On the Properties of Things*, II, p. 1290. Philip Ball, *Bright Earth: Art and the Invention of Color* (London: Penguin, 2001), p. 15.

greenness of this meadow removes what is turbid in the eyes and makes them bright and clear.[45]

With age, the body's natural heat gradually declined, along with its *humidum radicale*, or vital internal moisture, marking a shift from the balanced sanguine temperament towards one that was colder and drier, like the earth.[46] The eyes too suffered, losing their moisture, and becoming dehydrated and sore. This, in turn, caused irritation, failing sight and, eventually, blindness. Such phenomena were regarded as part of the natural process of ageing, occasioned by an inevitable change in humoral composition. It was, however, generally accepted that, although the individual could not escape this deterioration, through closely adhering to a *regimen* he or she could attempt to postpone it. Not surprisingly, the therapeutic use of colour figured prominently in these guides. For example, in his *De oculis*, Peter of Spain recommended constant exposure to green meadows as a means of preserving the sight.[47] Such ideas were not simply confined to an academic elite. Late medieval preachers clearly assumed a widespread awareness that 'grene colour makuþ men glad and bryngyþ counfort to þer y3en', and that gazing upon a meadow or stretch of grass would improve the crystalline humour of the eye, both strengthening and soothing it, as Albertus Magnus had suggested.[48]

Monks, in particular, were thought to benefit from contemplating green grass. One of the principal activities in the larger religious houses was the study and copying of religious and learned texts. Monks spent hours in the poor light of the *scriptorium* copying and illuminating substantial works, ironically often on medicine, such as the celebrated

---

[45] Hildegard of Bingen, *On Natural Philosophy and Medicine: Selections from Cause et cure*, trans. Margaret Berger (Cambridge: Brewer, 1999), p. 107.

[46] Vivian Nutton, 'Medicine in the Greek World, 800–50 BC', in *The Western Medical Tradition: 800 BC to AD 1800*, ed. Lawrence I. Conrad et al. (Cambridge: Cambridge University Press, 1995), pp. 11–38 (p. 25); Michael McVaugh, 'The *humidum radicale* in thirteenth-century medicine', *Traditio*, 30 (1974), 259–83; *Lydgate and Burgh's Secrees of Old Philisoffres*, ed. R. Steele, EETS, e.s. 66 (London: Keegan Paul, Trench & Trübner, 1894), p. 45.

[47] Daly and Yee, 'The Eye Book of Master Peter of Spain', p. 137; for a late fifteenth-century Latin copy of 'Experimentum Magistri Petri hispani contra omnem maculam oculorum', see Wellcome Institute Library, Western MS 618, fols 1–33ᵛ.

[48] *English Wycliffite Sermons*, ed. Anne Hudson, 5 vols (Oxford: Clarendon Press, 1983–1996), I, 603; V, 63.

eleventh-century herbal produced at Bury St Edmunds.[49] The constant strain had a detrimental effect on their eyesight. The author of one fifteenth-century leechbook makes the specific point that reading 'much of small letters' is 'evil for the sight',[50] confirming that many scholars and clerks suffered from eye strain after working long hours, especially by candle-light. When William Caxton completed his translation of *The Recuyell of the Historyes of Troye*, he commented that it was of such a length that his pen was worn and his eyes had become 'dimmed with ouermoche lokyng on the whit paper'.[51] This was not an uncommon grievance; the poet and civil servant, Thomas Hoccleve (d. c.1430), complained bitterly about the difficult conditions in which professional scribes had to work, enduring tired eyes and blurred vision, amongst many other ailments.[52] For those whose employment was reliant upon their excellent eyesight, using a salve to ease eyestrain would be the first option. One of the clerks in the service of Lord John Howard, later Duke of Norfolk, copied such a remedy into the household book for 1464, recommending that the reader should 'take a lytell white coperosse [copperas] and bray [pound] it and put it in a lytell rennynge watr, and putt it to the yhen'.[53] This recipe had obviously proved successful in alleviating the symptoms of painful and overworked eyes. We can thus appreciate one reason why the cloister garth, the central lawn around which the monks processed and studied, was so important to monastic communities. Monks could read their books in the cloisters, which

---

[49] Minta Collins, *Medieval Herbals: The Illustrative Traditions* (London: British Library, 2000), pp. 196–99.

[50] *A Leechbook or Collection of Medical Recipes of the Fifteenth Century*, ed. Warren R. Dawson (London: Macmillan, 1934), p. 161.

[51] *The Recuyell of the Historyes of Troye*, trans. William Caxton, ed. H. Sommer (London: D. Nutt, 1892). The belief that reading or writing on white paper damaged the eyes apparently dates back to Galen: *Galen On Sense Perception: His Doctrines, Observations and Experiments on Vision, Hearing, Smell, Taste, Touch and Pain, and their Historical Sources*, ed. Rudolph E. Siegel (New York: Karger, 1970), p. 81.

[52] *Hoccleve's Works: The Minor Poems including The Regement of Princes*, ed. Frederick J. Furnivall and I. Gollancz, EETS, e.s. 72 (London: Keegan Paul, Trench & Trübner, 1897), pp. 37–38.

[53] *The Household Books of John Howard, Duke of Norfolk, 1462–1471, 1481–1483*, ed. A. Crawford, 2 vols (Stroud: Sutton for Richard III and Yorkist History Trust, 1992), I, part I, p. 280.

provided good daylight and allowed them, periodically, to look out over the green grass to rest and refresh their weary eyes.[54]

Iconographical and archaeological evidence suggests that, from the mid-fourteenth century, long-sighted craftsmen, as well as monks and secular clergy, benefited from the development of convex lenses as reading glasses.[55] Fragments of spectacle frames have been found during excavations of monastic institutions in Britain, including Battle Abbey in West Sussex, the Dominican Friary at Chester and Melrose Abbey on the Scottish border.[56] Customs records for the latter part of the fourteenth century show that, even at this comparatively early date, glasses were being imported into England from the Continent in significant numbers: in just two and a half months, from 1 July to Michaelmas 1384, 160 pairs of spectacles were shipped into the port of London alone.[57] Amongst the purchasers who utilised these new reading aids were merchants, lawyers and ecclesiastics.[58] An inventory of the goods of the late Henry Bowet, archbishop of York (d. 1423), records a pair of eye-glasses with silver and gold frames valued at no less than 20s.[59] This was a particularly expensive and ornate pair, of the kind that only the affluent could afford. The book of expenses kept by the almoner of Peterborough abbey, however, shows that spectacles could be far less expensive; in 1463 he paid just 6d. to a local barber for supplying a pair.[60] The increasing availability of spectacles at a reasonable price proved an invaluable asset for long-sighted craftsmen and artisans struggling to continue their trade with deteriorating vision.[61]

[54] Sylvia Landsberg, *Medieval Gardens* (London: British Museum Press, 2002), p. 35. Carole Rawcliffe, "Delectable Sightes and Fragrant Smells': Gardens and Health in Late Medieval and Early Modern England', *Garden History*, 36 (2008), 3–21.

[55] Richard Corson, *Fashions in Eyeglasses* (London: Peter Owen, 1967); see frontispiece and figures 1–10.

[56] Judith Stevenson, 'A New Type of Late Medieval Spectacle Frame from the City of London', *London Archaeologist*, 7 (1995), 321–27 (p. 321).

[57] Michael Rhodes, 'A Pair of Fifteenth-Century Spectacle Frames From the City of London', *Antiquaries*, 62 (1982), 57–73 (pp. 59–60, especially n. 71).

[58] Rhodes, 'Fifteenth-Century Spectacle Frames', p. 65.

[59] *Testamenta Eboracensia: A Selection of Wills From The Registry at York*, ed. J. Rayne, Surtees Society, 45 (1864), p. 75.

[60] *The Book of William Morton, Almoner of Peterborough Monastery*, ed. W. T. Mellows, P. I. King and C. N. L. Brooke, Northamptonshire Record Society, 16 (1954), p. 164.

[61] Rhodes, 'Fifteenth-Century Spectacle Frames', pp. 64–65.

In this pre-Cartesian society, the immortal soul, whose health was regarded as being of far greater importance than that of the body, appeared to dwell within its inner recesses in close proximity to the animal spirits. Images reaching the brain through the eyes, therefore, could have a profound effect upon the soul as well as the spirits.[62] In *On Conversion*, Bernard of Clairvaux (c.1090–1153) describes the consequences for those who fail to protect their eyes from licentious or otherwise corrupting sights, thereby permitting spiritual pollution to enter the body. An uncontrolled roving eye could, potentially, transmit an overabundance of foul images to the brain, causing the memory to become an 'overflowing sewer', contaminating the whole body 'with intolerable filth'.[63] Bernard portrays the sense organs as being like windows and doors which must be closely guarded to prevent 'the entry of fresh filth' and to allow the individual 'to clean up what is already there'.[64] Only when the eye weeps rather than wanders and the sinner repents can the spiritual malaise be halted. The concept that sin could enter the body through the eye is a common trope in medieval homiletic literature and sermons.[65] For example, the author of the *Ancrene Riwle* warns anchoresses not to watch the world through their windows, because

> Of Eve, our first mother, it is written that sin entered first through her eyesight […] that is, Eve looked at the forbidden apple and saw that it was beautiful and began to take delight in looking at it and directed her desire to it and took it and ate of it and gave it to her lord.[66]

Instead, they were encouraged to cover their windows with thick black cloth which would shield the eyes from temptation.[67]

---

[62] Hildegard of Bingen, *Cause et cure*, p. 117; Adelard of Bath, 'Natural Questions', in *A Sourcebook in Medieval Science*, p. 377.

[63] Bernard of Clairvaux, *On Conversion*, in *Bernard of Clairvaux: Selected Works*, trans. Gillian R. Evans (New York: Paulist Press, 2005), p. 11.

[64] Bernard of Clairvaux, *On Conversion*, p. 12.

[65] See, for instance, Gerald of Wales, *The Jewel of the Church: A Translation of Gemma Ecclesiastica by Giraldus Cambrensis*, trans. J. Hagen (Leiden: Brill, 1979), pp. 182–83.

[66] *Ancrene Wisse: Guide for Anchoresses*, trans. Hugh White (Harmondsworth: Penguin, 1993), p. 29; also see *Ancrene Wisse: A Corrected Edition of the Text in Cambridge, Corpus Christi College, MS 402*, ed. Bella Millett, EETS, o.s. 325, 2 vols (Oxford: Oxford University Press, 2005), I, 21.

[67] *Ancrene Wisse: Guide for Anchoresses*, p. 28.

Just as external images could penetrate the body and exercise a profound effect on the soul, it also appeared that, when gazing into a person's eyes, one might observe his or her soul radiating outwards. Hildegard of Bingen maintained that 'the vision of the soul reveals itself in the eyes of a person [...] For a human's eyes are the windows of the soul'.[68] During the Middle Ages, the eyes, alongside other facial features, provided important signs for others to read. A person with clear, shining eyes was believed to have a radiant soul. Knowledge of this concept is assumed by Jocelin of Brakelond, a monk at the Abbey of Bury St Edmunds, who, when recording the election of a new abbot at the beginning of the thirteenth century, stressed that 'his eyes were crystal clear, with a penetrating gaze'.[69] From these few words alone, it is evident that the formidable Abbot Sampson is ideally suited for the role, as his soul appears to be without blemish, and he will be neither flattered nor fooled too easily. Clear eyes also denoted reliability and integrity. Guidance on what specific facial features revealed about a person's character was developed further by late medieval writers and whole tracts dedicated to the science of physiognomy quickly became popular. One example can be found in the fifteenth-century British Library MS Sloane 213. Its author states that he is providing 'certeyne rewles of phisnomy' so that his reader can know 'wen men lokes on any man of what condicions he es'.[70] It was commonly understood that the best-tempered individuals had 'blak eyghne, blak heres and roundenes of visage', whilst someone with 'eyghne like an asse' was thought to be 'a fole and of harde kynde [nature] and dulle.' Those whose 'eyghne [had] spottes, white, rede or blak' were best avoided altogether.[71]

It was very important to be able to look frankly and directly into another person's eyes, especially when conducting business transactions. As a result, someone who had 'perled' [cloudy] eyes ran the risk of appearing 'vntrew and intemperat'.[72] Not surprisingly, it was better to trade with a partner whose eyes were 'bright and clere' and who could maintain a steady gaze, as this was also a sign of probity and strength of

---

[68] *Hildegard of Bingen: Cause et cure*, p. 117.
[69] *Chronicles of the Abbey of Bury St Edmunds*, ed. Diana Greenway and J. Sayers (Oxford: Oxford University Press, 1989), p. 36.
[70] *Secretum secretorum*, p. 10.
[71] Ibid., pp. 10–11.
[72] Ibid., p. 98.

purpose.⁷³ The blind and pur-blind, whose eyes, and thus their characters, could often not be read, were regarded with suspicion; damaged or opaque eyes could be hiding a diseased soul. Not only were blind men and women often excluded from the medieval trading community but they were also unable to participate fully in the principal ecclesiastical ritual, the Mass. This was a predominantly visual experience for most medieval Christians as, under normal circumstances, they only expected to consume the Host once a year, usually at Easter.⁷⁴ The blind were also unable to benefit from the battery of uplifting images of Christ and his saints in their parish churches, which provided spiritual succour and inspiration to the sighted members of the congregation. It is understandable, therefore, that Bartholomaeus Anglicus believed that, compared with those suffering from other sensory impairments, blind men and women were the most 'wretched'. He states in his encyclopaedia that being trapped in the dark with little visual stimulation was comparable to being in 'a prisoun'.⁷⁵

The weekly Mass provided the congregation with the opportunity of witnessing the spectacle of transubstantiation. The climax of the ritual occurred when the bread and wine were transformed into the body and blood of Christ. Viewing the Host as it was raised high above the altar by the priest seemed akin to a dose of heavenly medicine; anyone who witnessed the elevation would reputedly enjoy protection from both sudden death and blindness for the remainder of that day.⁷⁶ Frequently, a bell would be rung to warn the congregation to look up in time from its prayers; the spectacle invigorated the vital spirits, thereby strengthening and nourishing the whole body.⁷⁷ The reserved sacrament was almost always illuminated by oil lamps or candles; a light shining over the altar indicated that the Host was present. Diocesan statutes from the early

---

[73] Ibid.; blind men of wealth often appointed attorneys to conduct transactions on their behalf to prevent discrimination and reduce the opportunities of being deceived: *Calendar of Patent Rolls, 1313–1317*, p. 472; *1317–1321*, p. 151.

[74] Bernard Lord Manning, *The People's Faith in the Time of Wyclif* (Cambridge: Cambridge University Press, 1919), pp. 61–62. Eamon Duffy observes that, 'for most people, most of the time the Host was something to be seen, not to be consumed' (*The Stripping of the Altars: Traditional Religion in England, c.1400–c.1580*, 2nd edn (New Haven, MA: Yale University Press, 2005), p. 95).

[75] *On the Properties of Things*, I, p. 364.

[76] John Myrc, *Instructions for Parish Priests*, ed. E. Peacock, EETS, o.s. 31 (London: Keegan Paul, Trench & Trübner, 1902), p. 10.

[77] Rawcliffe, *Leprosy*, p. 339.

thirteenth century stipulated that 'a lamp must burn continuously day and night before the host'.[78] In addition, it was commonplace for the raising of the 'transubstantiated blood and wine for adoration', to be 'accompanied by the raising of candles'.[79] The simultaneous elevation of the sacrament and candles served the dual purpose of illuminating the Host, increasing its visibility throughout the church, and reinforcing the Biblical associations between Christ, salvation and light.[80] Consequently, patrons bequeathing gifts to churches and hospitals in their wills often requested that lamps be lit in their memory to give succour to the weak and sickly by providing them with a constant source of comfort.[81]

Since the eyes provided such an essential but vulnerable mechanism for preserving both physical and spiritual health, it is understandable that anxiety about the loss of sight went beyond the obvious concerns that we in the West today would recognise. Medieval handbooks on health often gave instructions on how best to protect and strengthen the eyes, and hundreds of remedies designed to 'fortify the eyesight' may still be found in a variety of non-medical sources, including household accounts, musi-

---

[78] David Postles, 'Lights, Lamps and Layfolk: "Popular" Devotion Before the Black Death', *Journal of Medieval History*, 25 (1999), 97–114 (p. 99). See Statutes of Worcester 1240: *Councils and Synods with Other Documents Relating to the English Church*, ed. F. M. Powicke and C. R. Cheney, 2 vols (Oxford, 1964), II, 300.

[79] R. N. Swanson, *Religion and Devotion in Europe, c.1215–c.1515* (Cambridge: Cambridge University Press, 1995), p. 100.

[80] Duffy, *The Stripping of the Altars*, p. 96; D. R. Dendy, *The Use of Lights in Christian Worship* (London: SPCK, 1959).

[81] Eamon Duffy argues that the maintenance of ecclesiastical lights 'became the single most popular expression of piety in wills of the late medieval laity' (*The Stripping of the Altars*, p. 134). Specific examples of bequests by both laymen and clergy can be found in the *Calendar of Wills Proved and Enrolled in the Court of Husting, London, 1258–1688*, ed. Reginald R. Sharpe, 2 vols (London: Printed by John C. Francis, Took's Court, 1889–90). For example, in the plague year of 1349, William de Rothyng, a merchant of London, bequeathed 'to the Hospital of St Thomas in Southwark an annual quitrent for the maintenance of a lamp to burn by night among the weak and sickly there housed'. In the same year, the rector of the parish church of St Mary Magdalene, near the Old Fish Market, London, stipulated that rent from one of his tenements was 'to be charged with the maintenance of a lamp to burn before the Host in the chancel of the said church' (*Calendar of Wills*, I, 593, 595). In 1369, John de Potenhale likewise instructed that the revenue collected from his tenements should be used to provide 'a wax torch of twelve pounds weight to burn before the high altar [...] at the elevation of the Host' in St Mary's chapel in the church of St Andrew at Castle Baynard (*Calendar of Wills*, II, 133).

cal manuscripts and cookery books.[82] The fear of being unable to see the elevation of the Host or to benefit from the visual impact of sacred images in part explains why *regimina* provided so much guidance for the preservation of good eyesight as part of an everyday routine. For example, in his *Fifteen Directions to Preserve Health* (1602), William Vaughan advises the reader on rising in the morning to

> wash your face, eyes, eares, and hands, with fountaine water. I have knowne diuers students which vsed to bathe their eyes onely in well water twise a day, whereby they preserued their eyesight free from all passions and bloodsheds, and sharpened their memories maruaylously. You may sometimes bathe your eyes in rosewater, fennel water, or eyebright water.[83]

Here, the beneficial effects of pure water are clearly identified. It had been long understood that there were many different types of water, not all of which possessed therapeutic properties.[84] For instance, salt water could irritate and injure the eyes, whilst clean, flowing water, was believed to strengthen them.[85] Vaughan recommended fennel water as it was deemed to be relatively (but not excessively) warm and dry, and thus ideal for counteracting the phlegmatic (cold and moist) nature of the eyes. Although Vaughan's work was published at the turn of the sixteenth century, it underscores the importance placed on regularly cleansing the eyes, which was clearly recognised by medieval men and women. Fifteenth-century herbals frequently observe that, because fennel is 'hot and drie i[n] þe 2nd degree, it is gode for the eyen'.[86]

Herb-infused waters were commonly used to treat ophthalmic complaints, and instructions for their production often formed a substantial section in the specific genre of books on the diseases and

---

[82] *The Household Books of John Howard*, I, part I, p. 280; BL, Add. MS 10336, fol. ij; BL MS Sloane 442.

[83] William Vaughan, 'Fifteen Directions to Preserve Health', in *The Babees Book*, ed. Frederick J. Furnivall, EETS, e.s. 32 (London: Keegan Paul, Trench & Trübner, 1868), pp. 250–51. Also see Pedro Gil Sotres, 'The Regimens of Health', in *Western Medical Thought from Antiquity to the Middle Ages*, ed. M. D. Grmek, trans. Anthony Shugaar (Cambridge, MA: Harvard University Press, 1998), pp. 291–318 (p. 312).

[84] *Secrees of Old Philisoffres*, pp. 59–60; Avicenna, *The Canon of Medicine of Avicenna*, trans. O. Cameron Gruner (London: Luzac & Co., 1930), pp. 221–28.

[85] *The Medieval Health Handbook*, no. 107; Daly and Yee, 'The Eye Book of Master Peter of Spain', p. 137.

[86] BL, MS Sloane 963, fol. 138^r; compare with BL, MSS Sloane 3217, fol. 138^r; Lansdowne 680, fol. 17^v.

health of the eyes which became popular in the later Middle Ages.[87] Indeed, in his influential *De oculis*, Peter of Spain prescribed a 'miraculous water for preserving the vision', which included an amalgam of fennel, rue and euphrasy (eyebright) infused in white wine.[88] Keeping the eyes healthy and preventing a deterioration in vision concerned the medical elite and educated layman alike. Now bound in British Library MS 2276, *Grievances of the Eyes* is a typical example of what one might expect to find in a late medieval medical text dedicated to such a subject. It comprises more than fifty folios, listing and describing in great detail the different ailments which could affect the eyes. The inventory includes 'apostumes [a gathering of badly digested humours] [...] teris [tears], webbis [web-like growths across the eye] and catharactis [cataracts]'.[89] The anonymous author of *Grievances* also provides remedies ranging from surgery and cauterisation to less drastic alternatives, such as washing the eye with warm water and milk or using an eye ointment, referred to as a collyrium.[90] For instance, when treating cataracts, if the 'uniu[er]sell' treatments of 'diett and purgacion' had proved unsuccessful, it was suggested that the practitioner should first bathe the patient's eye with warm water and then anoint the cataract with an infusion of 'water of reede popy [red poppy] and centorie minor [lesser centaury] medlid [mixed] w[i]t[h] hony and [...] poligonia [knotweed]'.[91] Remedies such as these utilised a standard list of herbs, itemised by Peter of Spain in his *De oculis*, which included rue, celandine, fennel, wormwood, pimpernel

---

[87] It is possible that this genre derived from the work of the Muslim scholar and physician Hunain b. Ishaq (also known as Johannitius), notably his *Ten Treatises on the Eye*. Nancy Siraisi states that his work was 'especially influential in the medieval West [and] was translated into Latin before the end of the eleventh century' (*Medieval and Early Renaissance Medicine*, p. 108). Texts devoted to the eyes remained a discrete genre throughout the early modern period; see Richard Banister, *A Treatise of One Hundred and Thirteen Diseases of the Eyes and Eye-liddes* (London: Imprinted by Felix Kyngston, for Thomas Man, 1622).

[88] Daly and Yee, 'The Eye Book of Master Peter of Spain', p. 135.

[89] BL, MS Sloane 2276, fol. 164ʳ. Definitions can be found in Getz, *Healing and Society*, pp. 311–361.

[90] Cautery involved using a hot iron to remove diseased skin, thereby eliminating excessive humours associated with it. See Jones, *Medieval Medical Miniatures*, p. 97; for examples, see BL, MS Sloane 2276, fols 173ʳ–175ʳ; for recommendations of washing and ointments, see fols 166ᵛ, 169ᵛ.

[91] BL, MS Sloane 2276, fol. 182ʳ.

and vervain.[92] In order for them to be made into a plaster or an ointment that could be easily applied to the eye, a binding agent, such as honey, was needed. Egg white was also popular as a base substance for making ointments; its cooling properties made it ideal for use after any surgical treatment, a practice advocated by the author of *Grievances*.[93] Furthermore, albumen shared the same composition as the phlegmatic eye, since both were inherently cold and moist; it was, thus, believed to be more efficacious in promoting recovery.[94]

Remedies for ophthalmic complaints caused or exacerbated by old age, such as glaucoma, cataracts and rheumy eyes, are frequently included these books. It was recognised that with age eyesight dims. In his *De oculis*, Peter of Spain recommends an unguent 'for preserving and improving vision in age' made from rabbit bile, honey and morning dew, ingredients specifically chosen for their soothing and moistening properties.[95] John of Arderne (b. 1307) was another practitioner with an interest in treating the ailments caused by old age, especially the ubiquitous problem of watering eyes. A skilled botanist and renowned surgeon, Arderne gained fame in England for his treatment of, and celebrated work on, anal fistulae, but he also composed a tract entitled *De cura oculorum*. Creating an ointment from ingredients as varied as man's urine, capon fat and butter, Arderne maintained that, until the age of seventy, he had used the medicine himself and it had proved highly beneficial, alleviating the symptoms caused by close work in poor light.[96] Both Arderne and Peter of Spain composed their work on the eye in the later years of their careers. It is understandable that they were most concerned about ophthalmic conditions at a time of life when they, too, possibly suffered from age-related problems aggravated by long hours studying and reading.[97]

---

[92] Daly and Yee, 'The Eye Book of Master Peter of Spain', p. 137.

[93] BL, MS Sloane 2276, fols 169^v, 174^v. It is now known that egg white contains 'lysozyme, a natural enzyme found in tears, which has mild antibiotic properties': *Benvenutus Grassus: The Wonderful Art of the Eye: A Critical Edition of De probatissima arte oculorum*, ed. L.M. Elderedge (East Lansing, MI: Michigan State University Press, 1996), p. 20.

[94] BL, MS Lansdowne 680, fols 7^r–9^r.

[95] Daly and Yee, 'The Eye Book of Master Peter of Spain', p. 140.

[96] BL, MS Sloane 75, fol. 182^v.

[97] Daly and Yee, 'The Eye Book of Master Peter of Spain', p. 123.

We have seen that the eye was considered to be one of the noblest organs in the body and sight the most precious of the senses; consequently, the blind seemed incomplete and severely disadvantaged. Indeed, Bartholomaeus Anglicus described blindness as being like a prison, enclosing the individual within a private world of darkness. Although those with a visual impediment were often firmly integrated into many aspects of contemporary social and religious life, a lack of sight clearly separated the blind from their seeing peers. Not only did the quality of the eyes reflect a person's general fitness and character, but also acted as a window, revealing the deepest secrets of the soul. Thus the authors of physiognomy texts explained that the blind could never be fully trusted because opaque eyes could be disguising a malign temperament. Moreover, those without vision were unable to benefit from the therapeutic effects of a variety of stimuli, such as those transmitted by jewels, green grass and religious iconography, which were believed to fortify the vital and animal spirits. It is not surprising, therefore, that medieval men and women were prepared to go to great lengths to protect and restore their eyesight, thereby maintaining their physical and spiritual well-being.

Illuminated Darkness:
The Image of the Blind or Blindfold Man in Some Thirteenth- and Fourteenth-Century European Manuscripts

Sabina Zonno

On the left lower edge of a miniature at the beginning of a sumptuous little-known psalter illuminated in East Anglia between about 1320 and 1330, and now kept in Brescia (Biblioteca Queriniana, MS A.V.17, fol. 7^r), a man with his eyes closed and a staff in his left hand — in the usual iconography of the blind man — is led with a rope by a child (fig. 1). He is at the rear of a procession of destitute people including the child just mentioned, a bald man who may be possibly identified with a fool,[1] an infant carried in back slings by his mother, another, less visible woman, and a bearded man with a walking stick. They are all standing outside the divine court of king David portrayed as a medieval sovereign in a magnificent towered castle with blue roofs and domes, doors and parapets, inhabited by twelve noble onlookers listening to the harmonious music of his harp. An official of David's court at the open gate seems to drive the blind, the poor, and the needy from the door. Moving from the assumption that illuminations are vivid expressions of medieval culture and provide a means for understanding our past, contributing thus to explain our present, I have investigated the iconographic tradition of the blind man – as blind from birth, momentarily blind, or blindfolded – in a selection of sacred and lay codices illuminated in Europe in the same period as the Brescia Psalter. The aim is to make this unconventional iconography clear and study the perception of blindness in the Middle Ages, so as to retrace the possible origins of a prejudicial treatment of blind people. These illuminations will be analysed in their specific context and in relation to the text they

---

[1] D. J. Gifford, 'Iconographical Notes towards a Definition of the Medieval Fool', *Journal of the Warburg and Courtauld Institutes*, 37 (1974), 336–343; Marco Assirelli, 'L'immagine dello "stolto" nel salmo 52', in *Il Codice miniato: rapporti tra codice, testo e figurazione, Atti del III Congresso di Storia della Miniatura, Cortona, 20–23 ottobre, 1988*, ed. M. Ceccanti and M.C. Castelli (Firenze: L.S. Olschki 1992), pp. 19–27.

accompany exploring the possible exegetic and symbolical meanings connected with the image of blindness.

Not surprisingly, the concept of blindness occurs copiously in the Bible, both in the Old and the New Testament, and in the patristic interpretations of the Holy Scriptures too. It is very common to find in the biblical passages some references to blindness associated with the traditional dichotomy between light and darkness. The contrast mirrors the perennial conflict between divine illumination and diabolic obscurity: the first is evidently connected with virtue and sight, while the second is linked to sin and blindness.[2] It is in the New Testament that Christ Himself states that He is the light,[3] and this statement is corroborated by the illuminations. In a Gothic North-Eastern French Psalter produced in the mid-thirteenth century and now in Douai (Bibliothèque Municipale, MS 186, fol. 39^v),[4] this idea concretises in the portrait of David raising a candle as a symbol of light toward the Lord who is depicted above an altar. In the psalter tradition where Psalm 26 reads 'The Lord is my light and my salvation', king David is usually shown pointing to his eyes before Christ within the historiated initial heading the psalm.[5] The image visually translates *ad verbum* the first verse of the prayer[6] and it also appears in the Brescia Psalter where David is depicted wrapped in a blue baggy mantle kneeling down between two cliffs; he indicates his eyes turning his head toward the Lord who appears behind a curtain of clouds

---

[2] For the emblematic relations between sight and faith, and blindness and unbelief, see F. Dorothy Glass, '"Then the eyes of the blind shall be opened": A Lintel from San Cassiano a Settimo', in *Reading Medieval Images: The Art Historian and the Object*, ed. E. Sears and Th. K. Thomas (Ann Arbor: University of Michigan Press, 2002), pp. 142–150.

[3] From the Gospel of John: 'Again Jesus spoke to them, saying, I am the light of the world. Whoever follows me will not walk in darkness, but will have the light of life.' (8. 12). References to the Bible are to the Douay-Rheims translation.

[4] See *Enluminures*: <http://www.enluminures.culture.fr/documentation/enlumine/fr/index3.html> [accessed 16 January 2013].

[5] See for example Günther Haseloff, *Die Psalterillustration im 13. Jahrhundert: Studien zur Geschichte der Buchmalerei in England, Frankreich und den Niederlanden* (Kiel: Selbstverlag des Verfassers, 1938), pp. 24–71; Frank O. Büttner, 'Der illuminierte Psalter im Westen', in *The Illuminated Psalter: Studies in the Content, Purpose and Placement of its Images*, ed. F.O. Büttner (Turnhout: Brepols, 2004), pp. 1–106 (pp. 19–21).

[6] See Lucy Freeman Sandler, 'The Images of Words in English Gothic Psalters', in *Studies of the Illustration of the Psalter*, ed. B. Cassidy and R. Muir Wright (Stamford: Shaun Tyas, 2000), pp. 52–68 (pp. 67–86).

above. With this gesture, David emblematically suggests that both his light and sight originate in God. So, he embodies the right and virtuous man illuminated by divine light.

But a major step forward in the explanation of the symbolical connotation of light and consequently in the explanation of blindness is taken by Adelaide Bennett who has recently observed that 'the gesture of pointing to the eye signifies comprehension of illumination and hope for salvation'.[7] It is noteworthy in this sense that in the sacred writings Christ appears as the man who can illuminate darkness restoring sight to those people who are unable to see. The standard illumination portraying David who points to his eyes is sometimes replaced by the healing of a blind man in Psalm 26.[8] This miracle is depicted for example in the historiated initials in a group of psalters and books of hours that have been produced in Liège from the second half of the thirteenth century. This episode refers to the Gospel of Luke (18. 35-53) where the Evangelist tells the story of Christ who healed a blind beggar.[9]

In medieval illuminations, sightless men are commonly represented as solitary travellers or mendicants with a staff in their hand, as in the Brescia Psalter. They have a hat or cap on their head and a small shoulder bag as in the Flemish exemplars taken into consideration. As noted in the Gospel of St John,[10] blind people do not know where to go; consequently, someone or something must guide them: another man or a child can show them the way, as in the Brescia miniature, or simply a dog as in another psalter now in Liège (University Library, MS 431, fol. 31ᵛ).[11] Blindness can thus be cured by Christ with the sign of a cross on

---

[7] Adelaide Bennett, 'The Transformation of the Gothic Psalter in Thirteenth-Century France', in *The Illuminated Psalter*, ed. F.O. Büttner, pp. 211–221 (p. 215).

[8] Judith H. Oliver, *Gothic Manuscript Illumination in the Diocese of Liège (c. 1250–c.1330)* (Leuven: Uitgeverij Peeters, 1988), p. 65. On the iconography of the healing of the blind man, see Barbara Baert, 'The Healing of the Blind Man at Siloam, Jerusalem: A Contribution to the Relationship Between Holy Places and the Visual Arts in the Middle Ages (Parte I)', *Arte cristiana*, 95 (fasc. 838) (2007), 49–60; Barbara Baert, 'The Healing of the Blind Man at Siloam, Jerusalem: A Contribution to the Relationship Between Holy Places and the Visual Arts in the Middle Ages (Parte II)', *Arte cristiana*, 95 (fasc. 839) (2007), 121–130.

[9] Judith H. Oliver, p. 65.

[10] From the Gospel of John: 'Jesus therefore said to them: Yet a little while, the light is among you. Walk whilst you have the light, that the darkness overtake you not. And he that walketh in darkness, knoweth not whither he goeth' (12. 35).

[11] This image is published in Oliver, p. 374, for example.

the forehead,[12] as documented in a book of Psalms produced in Flanders between 1290 and 1305 and now in New York (Pierpont Morgan Library (PML), MS M.155, fol. 22ᵛ),[13] or by His miraculous touch as shown by a coeval psalter now in Brussels (Koninklijke Bibliotheek, MS IV-141, fol. 32ʳ).[14] The miracle He performs is the symbol of the spiritual blindness which He comes to earth to redeem. This idea is substantiated, for example, by the presence of Moses in the bas-de-page of Psalm 26 of a Psalter-Book of Hours now in The Hague (Koninklijke Bibliotheek, MS 76 G 17, fol. 20ᵛ)[15] and illuminated at the end of the thirteenth century for the use of Liège. While Christ is healing a man born blind within the historiated initial, in the lower margin of the page Moses exhibits a scroll with the words 'Veniet illuminator tenebris'; thus, he announces the Messiah's arrival and proclaims the role of Christ as illuminator of the world. These illuminations – like the ivory diptychs with the same subject – show the path by which sight can be restored, the 'sight both of the body and of the soul' (Singelenberg, 1958, 107). In the Taymouth Book of Hours (London, British Library (BL), Yates Thompson MS 13, fol. 103ʳ)[16] executed in England between 1325 and 1335,[17] Christ is depicted in front of His disciples healing a blind man with his eyes tightly closed in the margin of Psalm 125 (fig. 2). Intriguingly, the miracle appears exactly under the word *Convertere* (Psalm, 125. 4) implying a direct connection between conversion to Christian faith and sight. In this way, Christ attests to his power to save blind people from their spiritual darkness. For example, the Epistle to the Hebrews suggests that mankind

---

[12] On the healing by a touch as a Christian concept, see Pieter J. Lalleman, 'Healing by a Mere Touch as a Christian Concept', *Tyndale Bulletin*, 48.2 (1997), 355–361.

[13] See *Corsair: The Online Research Resource of the Pierpont Morgan Library* <http://utu.morganlibrary.org/medren/ListOfMssWithImages.cfm> [accessed 16 January 2013].

[14] The image is in Oliver, p. 375.

[15] See *Medieval Illuminated Manuscripts* <http://manuscripts.kb.nl/show/images/76+G+17> [accessed 16 January 2013].

[16] See the *Catalogue of Illuminated Manuscripts* <http://www.bl.uk/catalogues/illuminatedmanuscripts/record.asp?MSID=8148&CollID=58&NStart=13> [accessed 16 January 2013].

[17] Lucy Freeman Sandler, *Gothic Manuscripts 1285–1385: A Survey of Manuscripts Illuminated in the British Isles* (London-Oxford: Harvey Miller Publishers-Oxford University Press, 1986), pp. 107–109.

can obtain salvation and regain sight just through their conversion.[18] And in the Epistle to the Ephesians, the relation between sight and revelation is particularly emphasised: 'That the God of our Lord Jesus Christ, the Father of glory, may give unto you the spirit of wisdom and of revelation, in the knowledge of him', and 'The eyes of your heart enlightened' (1. 17-18). The illumination in the lower margin of the Taymouth Hours signifies Christ's capacity to restore sight and it reveals to the medieval reader that those people who follow Christ will not walk in darkness and their sight will be healed by faith. Respecting His laws, they will be illuminated by the bright light of Christ.

But Christ is not the only one who can cure people of their blindness. Even His disciples and messengers can perform this miracle in the name of God. In the Book of Tobit, for example, the virtuous protagonist of the story is made temporarily blind by the excrements of some birds (2. 11)[19] and it is the archangel Raphael who appears to his son Tobias to restore Tobit's sight with his help (11. 1-21). In thirteenth- and fourteenth-century biblical iconography, Tobit is usually shown alone lying on his bed with his eyes closed and the swallows flying above him. However, in an initial of a Bible in New York (PML, MS M.138, fol. 151^v)[20] possibly illuminated in England between the 1260s and the 1270s, Tobit is depicted with the archangel who is going to accompany his son in his trip. They are discussing and Raphael seems to be instructing the blind Tobit with his raised right forefinger. By contrast, in another Bible in New York (New York, PML, MS G.42, fol. 142^v)[21] executed in Oxford in the 1260s, the role of the archangel as a spiritual leader is particularly stressed: he holds the hand of Tobias as a child before his father, who is seated on a bench with his dog at his feet. It is noteworthy that before Tobit has his sight restored, the archangel says to his son Tobias: 'For be assured that his eyes shall be presently opened, and thy father shall see the light of heaven, and shall rejoice in the sight of

---

[18] 'But call to mind the former days, wherein, being illuminated, you endured a great fight of afflictions' (Hebrews 10. 32).

[19] The text reads 'And as he was sleeping, hot dung out of a swallow's nest fell upon his eyes, and he was made blind. Now this trial the Lord therefore permitted to happen to him, that an example might be given to posterity of his patience, as also of holy Job' (Tobit 2. 11–12).

[20] See *Corsair* <http://utu.morganlibrary.org/medren/single_image2.cfm?imagename=m138.151v.jpg&page=ICA000106538> [accessed 16 January 2013].

[21] See *Corsair* <http://utu.morganlibrary.org/medren/single_image2.cfm?imagename=g42.142v.jpg&page=ICA000122839> [accessed 16 January 2013].

thee' (Tobit 11. 8). Again, the connection between sight and divine light is clearly in evidence.

In the story of the conversion of Saul on his journey to Damascus, blindness seems to represent the transition from scepticism to belief and it is Ananias as a Christian disciple who has a fundamental role in the healing of the protagonist. Saul was a Jew by birth and a persecutor of the Christians and changed into a follower of Christ through revelation manifested as 'a light from heaven shined round about him' (Acts 9. 1–9). In the illustrative tradition of the Pauline Epistles, Saul is shown as a temporarily blind man who is about to fall down from a horse, lying face upward on the ground with his eyes shut, on the bed as a sick man, or led forward by a disciple who precedes him in the epistles of the Colossians, Corinthians, Galatians, Ephesians, Philippians, and Thessalonians.[22] Ananias restores his sight and states:

> And Ananias went his way, and entered into the house. And laying his hands upon him, he said: Brother Saul, the Lord Jesus hath sent me, he that appeared to thee in the way as thou camest; that thou mayest receive thy sight, and be filled with the Holy Ghost. And immediately there fell from his eyes as it were scales, and he received his sight; and rising up, he was baptized (Acts 9. 17–18).

The mystical experience of Saul consists in his momentary blindness and the miracle performed by Ananias can be interpreted as a type of salvation too.[23] Conversion to Christian faith is thus exemplified here by a change from blindness to sight and the association between the healing of Saul and baptism is highlighted by the biblical contents. Intriguingly, in a French exemplar illuminated in the 1250s and now in Vatican City (Biblioteca Apostolica Vaticana, Urb. MS lat. 7, fol. 390^r), Saul is portrayed not as a blind but as a blindfolded man. We may thus suppose that the blindfold is possibly used to convey the idea of a nonpermanent disease in this case. It is in fact a form of obstruction that makes people sightless for only a limited period of time. In any case, it can be removed or raised.

In some cases, even saints have a thaumaturgical power and restore the sight of blind people. But in the story of St Daniel who was probably less known to a non-Paduan public, the patron saint of the city was elected as

---

[22] For the iconography of the Pauline epistles, see Luba Eleen, *The Illustration of the Pauline Epistles in French and English Bibles of the Twelfth and Thirteenth Centuries* (Oxford: Clarendon Press, 1982).

[23] Ibid., p. 88.

messenger of Christ to persuade a blind man from Tuscany to leave for Padua and visit the Oratory of St Prosdocimo where his tomb was — though unbeknownst — to have his sight restored.[24] The hagiographical legend is summarised into a precious illuminated initial heading the feast of St Daniel in an Antiphonary in Padua (Biblioteca Capitolare, MS B.14, fol. 177^r), illuminated in the 1320s (Fig. 3).[25] In the initial, Daniel as a deacon appears above in a dream to the sightless man sleeping in his bedroom; below, the blind man is led by a young travelling companion, and they have both long walking sticks in the traditional iconography. Emblematically, his way to Padua could represent his religious path to faith and the miraculous healing can be considered the sign of the very favourable consequences brought by devotion on the one hand and a symbol of the sanctity of Daniel as an emissary of Christ on the other.

The relation between sight and religious worship is also implied in the Gospel of Matthew by the words of Christ after His miracle of the healing of two men born blind described in two distinct passages; in the first (9. 27–31), the Evangelist reports that Christ touched and opened the sightless eyes of those men who followed Him crying and saying, 'Have mercy on us, O Son of David' (9. 27); He finally said: 'According to your faith, be it done unto you' (9. 29); in the second (20. 30–34), Matthew reveals that after the miracle the men 'immediately [...] saw, and followed him' (20. 34). This episode gives an effective example of the complete trust in God and shows that their lack of sight is cured as a result of their devotion. A possible reference to this miracle can be seen in the well known *Omne Bonum* in London (BL, Royal MS 6. E. VI, fol. 245^r)[26] – a general encyclopedia written and illustrated in England during the third quarter of the fourteenth century – where in the initial C of *Cecus sive Cecitas* the meaning of blindness is explained by the scene of

---

[24] Ireneo Daniele, 'Daniele di Padova', in *Bibliotheca Sanctorum*, 13 vols (Rome: Istituto Giovanni XXIII della Pontificia Università Lateranense, 1964), IV, 474–475.

[25] Claudio Bellinati, 'Antifonario', in *Parole dipinte: la miniatura a Padova dal Medioevo al Settecento*, ed. G. Baldissin Molli, G. Canova Mariani, and F. Toniolo (Modena: Franco Cosimo Panini, 1999), pp. 89–90; Silvio Bernardinello, *Catalogo dei codici della Biblioteca Capitolare di Padova*, 2 vols (Padova: Istituto per la Storia Ecclesiastica Padovana, 2007), I, 179–180.

[26] See *Catalogue of Illuminated Manuscripts* <http://prodigi.bl.uk/illcat/ILLUMIN.ASP?Size=mid&IllID=32101> [accessed 16 January 2013].

two blind men kneeling in front of Christ accompanied by one of His disciples.[27]

In the Gospel of John (9. 1-41), the Healing of a blind from birth is even more detailed:

> When he had said these things, he spat on the ground, and made clay of the spittle, and spread the clay on his eyes, And said to him: Go, wash in the pool of Siloe, which is interpreted, Sent. He went therefore, and washed, and he came seeing (9. 6–7).[28]

A full-page miniature in the Psalter-Hours of Yolande de Soissons in New York (New York, PML, MS M.729, fol. 104ᵛ)[29] illuminated in Amiens at the end of the thirteenth century condenses the sacred writings into a single efficacious image.[30] The man who was blind is stepping from the pool of Siloam.[31] His eyes are open since he has already gained sight. The miniature portrays the moment after the miracle, when the man is asked questions of the miraculous event by the Pharisees depicted on the left. They refuse to believe in such a supernatural intervention and accuse the man of lying, implying also that his saviour is a sinner.[32] But the man replies bravely with his left forefinger lifted and defends Christ trying to persuade the Pharisees that only a man of God can restore sight. In the miniature the main protagonists of the story appear on the stage, but Christ is not present. As Alexa Sand underlines, 'the formerly blind man

---

[27] Lucy Freeman Sandler, *Omne bonum: A Fourteenth-Century Encyclopedia of Universal Knowledge. British Library MSS Royal 6 E VI – 6 E VII* (London: Harvey Miller Publishers, 1996).

[28] The miracle is described also by Mark (10. 46–52) and Luke (18. 35–43).

[29] See Corsair <http://utu.morganlibrary.org/medren/single_image2.cfm?imagename=m729.104v.jpg&page=ICA000001155 > [accessed 16 January 2013].

[30] See Alison Stones, 'The Full-Page Miniatures of the Psalter-Hours New York, PML, MS M.729. Programme and Patron', in *The Illuminated Psalter*, ed. F.O. Büttner, pp. 281–307; François Avril, 'Psautier-livre d'heures de Yolande de Soissons', in *L'art au temps des rois maudits Philippe le Bel et ses fils 1285–1328*, exhibition catalogue (Paris: Editions de la Réunion des musées nationaux, 1998), pp. 298–300.

[31] On the origins of this iconography, see Pieter Singelenberg, 'The Iconography of the Etschmiadzin Diptych and the Healing of the Blind Man at Siloe', *The Art Bulletin*, 40.2 (1958), 105–112; Baert, 49–60 and 121–144.

[32] 'Some therefore of the Pharisees said: This man is not of God, who keepeth not the sabbath. But others said: How can a man that is a sinner do such miracles? And there was a division among them' (John, 9. 16). 'They therefore called the man again that had been blind, and said to him: Give glory to God. We know that this man is a sinner' (9. 24).

has taken up the role of Christ, rising reborn to spiritual as well as literal sight from the water [...] Here is a powerful exemplum for the viewer: faith is a form of Christomimesis so efficacious that it raises one from darkness and death to light and true life, from blindness to vision'.[33] As Christ Himself reveals at the beginning of the Gospel of John: 'the works of God should be made manifest in him' (9. 3). Interestingly, St Ambrose observes that the Healing of the blind man deals with sight and insight since the man is enlightened in the doctrine of salvation.[34] Hence, the blind man becomes an example for the community who is witness to his metamorphosis from blindness to truth.

The same biblical passage from the Gospel of John introduces another stimulating topic related to blindness and its symbolic connotations. When Christ sees the blind man begging, His disciples ask Him: 'Rabbi, who hath sinned, this man, or his parents, that he should be born blind?' (9. 2). Such a question would seem to imply that blindness was thought to have its origins in wickedness. The Scriptures and the patristic interpretations of the sacred writings have certainly emerged as the main sources of inspiration for the symbolical and allegorical overtones of blindness. As Isidor of Sevilla (*ca.*560–633) points out, humanity was made sightless after the original sin and the following expulsion of Adam and Eve from Eden.[35] Consequently, the loss of devotion to God was considered the main source of human blindness. If the faculty of seeing is associated with divine light in the sources analysed, blindness appears to be related to sin. That is the other side of the coin, as shown in the Holy Scriptures. In another section of the Gospel of John, for example, the Evangelist reveals that there are men who 'loved darkness rather than the light: for their works were evil' (3. 19). These people who live in the dark can be possibly identified with those who do not respect Christian teachings. In the Second Epistle of Peter, the man who has no faith, virtue, knowledge, self-control, fortitude, or other good qualities, is said to be 'blind, and groping, having forgotten that he was purged from his old sins' (II Peter 1. 9). Hence, blindness mirrors all the moral defects caused by the inability to penetrate Christian beliefs.

In medieval society, blindness was thus conceived of both as a physical deformity and an image of spiritual imperfection that prevents people

---

[33] Alexa Sand, 'Vision, Devotion, and Difficulty in the Psalter Hours "Of Yolande of Soissons"', *The Art Bulletin*, 87.1 (2005), 6–23 (p. 16).

[34] Baert, 53.

[35] Glass, 142–150.

from seeing the material world in which God manifests Himself, precluding them from glimpsing the spiritual world too. This infirmity identifies those people who are considered responsible for their immorality, since 'a person's manner of gazing, his eye, and his sight were all indicators of the salvatory state of a person'.[36] As Job asserts, 'They have been rebellious to the light, they have not known his ways, neither have they returned by his paths' (24. 13). Like the destitute, the humble, and the miserable, sightless men were evidently deemed to merit their disease because misfortune was traditionally thought to derive from the Devil.[37] So, blindness could be also a divine punishment as the words of the prophet Zephaniah seem to suggest: 'And I will distress men, and they shall walk like blind men, because they have sinned against the Lord' (1. 17).

Blindness is thus an emblem of vice, depravity, and corruption. As a result, in the Middle Ages, blind people were marked as vulgar outcasts, being rejected by society and obliged to live in exile; they were regarded as inferior, foolish, and filthy, not only because of their social status, but for their inner moral condition, and the resultant negative judgment of God upon them.[38]

These considerations help us to investigate the presence of the blind man in the miniature of the Brescia Psalter and its meanings from which the discourse on blindness started (fig. 1). This image opens the book introducing the text of Psalm 1 in relation to which the image must be seen. The first verses read:

> Blessed is the man who hath not walked in the counsel of the ungodly, nor stood in the way of sinners, nor sat in the chair of pestilence. But his will is in the law of the Lord, and on his law he shall meditate day and night. [...] Not so the wicked, not so: but like the dust, which the wind driveth from the face of the earth. Therefore the wicked shall not rise again in judgement:

---

[36] Thomas Lentes, '"As far as I can see...": Rituals of Gazing in the Late Middle Ages', in *The Mind's Eye: Art and Theological Argument in the Middle Ages*, ed. J. Hamburger and A.-M. Bouché (Princeton: University Press, 2004), pp. 360–373 (p. 362).

[37] Aron Ja. Gurevič, *Le categorie della cultura medievale* trans. Clara Castelli (Torino: G. Einaudi, 1983), p. 9; Bronislaw Geremek, *La stirpe di Caino: l'immagine dei vagabondi e dei poveri nelle letterature europee dal XV al XVII secolo*, ed. F.M. Cataluccio (Milano: Il saggiatore, 1988), p. 7; Michel Mollat, *Les pauvres au Moyen Âge: Étude sociale* (Paris: Hachette, 1978), p. 4.

[38] See the fundamental work of Edward Wheatley, *Stumbling Blocks Before the Blind: Medieval Constructions of a Disability* (Ann Arbor: University of Michigan Press, 2010).

nor sinners in the council of the just. For the Lord knoweth the way of the just: and the way of the wicked shall perish (1. 1-6).

Here, the emblematic juxtaposition between virtue and sin is accentuated by the repetition of the words 'law' and 'just' versus 'wicked' and 'sinners'.[39] The same opposition seems to be emphasised by the antagonism between the holy court of king David and the vile crowd of destitutes outside the castle, including the blind man led with a rope by the child.[40] The presence at the open gate of the castle of the official of the court of David who seems to use his staff to drive this deplorable group away from the palace would seem to suggest the exclusion of these iniquitous people from the divine residence.[41] David, who is generally considered a very polyhedric biblical character, is exalted here as the author of the psalms and praised as a model of kingship and rectitude.[42] He personifies the illuminated medieval sovereign who knows the way of the worthy and lives in the congregation of the righteous.[43] And who are the wicked people walking in the counsel of the ungodly referred to in the text? Can the needy outside the castle typify the ungodly standing in the way of sinners? If we take into consideration now all the previous reflections on blindness and its negative connotations, we could agree with the idea that in the Brescia Psalter the blind man and his companions are not reminiscent just of 'the medieval social category of the wayfarers':[44] those impoverished people were 'driven to the road by urban or rural dispossession' in thirteenth- and fourteenth-century England because of the cost of economic growth, increasing prices and

---

[39] On the distinction between the virtuous and the wicked, see also Celia Chazelle, 'Violence and the Virtuous Ruler in the Utrecht Psalter', in *The Illuminated Psalter*, ed. F.O. Büttner, pp. 337–348 (p. 341).

[40] Sabina Zonno, 'Illumination Translates: The Image of the Castle in Some Fourteenth-Century English Manuscripts', in *The Medieval Translator-Traduire au Moyen Age: Vol. 12 Lost in Translation*, ed. D. Renevey and C. Whitehead (Turnhout: Brepols, 2009), pp. 297–313 (pp. 304–307).

[41] I am indebted to Professor V.A. Kolve of the University of California, Los Angeles for his help in interpreting this scene.

[42] Zonno, p. 304.

[43] Sabina Zonno, 'Il mirabile castello: Davide e la sua corte nel Salterio inglese della Biblioteca Queriniana di Brescia', *Annali Queriniani*, 6 (2005), 71–100 (pp. 95–96).

[44] Jean J. Jusserand, *English Wayfaring Life in the Middle Ages* (New York: Ernest Benn Limited, 1950).

taxes, and food shortages caused by severe winters[45] and obliged to beg for food and money at churches, abbeys, cathedrals, or castles where someone could show mercy feeding and clothing them, and giving them something to drink, as Matthew suggests in his Gospel.[46] The miniature would seem to show that king David is the *beatus* 'saved by God's grace [...], therefore, in the afterlife, will be in His presence for the rest of eternity, blissfully enjoying the Beatific Vision, the most spiritual of visions';[47] by contrast, the blind man with his companions are condemned to their wicked condition. In particular, the sightless man is doomed to his physical infirmity that mirrors his blindness to God. He does not merit David's compassion because of his immorality. Consequently, he is not welcome in the castle by reason of his blindness that prevents him from seeing God and His reign whose integrity and harmony must be necessarily defended by the king or by his officials. The exclusion of the blind man and these vicious people can possibly be read in relation to the Revelation where in the Holy Jerusalem 'there shall not enter anything that defiles or works abomination' (21, 27).[48]

My impression is that in the Middle Ages, blindness was conceived of as a particular condition limiting man's life but also marking him socially as an immoral outcast. But also the blindfold in medieval illuminations can be read as an emblem of the darkness of sin and a symbol of spiritual blindness. Unlike the permanent loss of sight, the blindfold represents a temporary state as we have seen in the discussion on the iconography of Saul. For example, the personification of the blindfolded Synagogue opposing the allegorical sighted female figure of the Church is a very popular motif in medieval art.[49] This anthitetical pair shows the

---

[45] Lucy Freeman Sandler, 'Pictorial and Verbal Play in the Margins: The Case of British Library, Stowe MS 49', in *Illuminating The Book: Makers and Interpreters. Essays in Honour of Janet Backhouse*, ed. Michelle P. Brown and Scot McKendrick (London: The British Library, 1998), pp. 52–68 (p. 53).

[46] 'For when I was hungry, you gave me food; when thirsty, you gave me drink; when I was a stranger you took me into your home, when naked you clothed me' (Matthew 25. 35–36). See also Proverbs 25. 21; Ecclesiastes 4. 1–11; Isaiah 58. 7; Zechariah 7. 9–10; Matthew 6. 3–4; 10. 40–42; Romans 12. 13.

[47] Katherine H. Tachau, 'Seeing as Action and Passion in the Thirteenth and Fourteenth Centuries', in *The Mind's Eye*, ed. J. Hamburger and A.-M. Bouché, pp. 336–359 (p. 347).

[48] Zonno, pp. 304–305.

[49] See for example Moshe Barasch, *Blindness: The History of A Mental Image in Western Thought* (New York: Routledge, 2001), pp. 79–91; Lilian M.C. Randall, *Images in*

victorious Church and the defeated Synagogue as an image allegorically referring to the final triumph of Christianity.[50] The blindfold alludes possibly to Moses's veil preventing the Jews from perceiving the true figurative glory of their text in Christ[51] and their consequent inability to view the light of redemption and the divinity of Christ. In the New Testament — as Barasch points out — 'the veil is explicitly transformed into a metaphorical expression of blindness'.[52] In the Spanish *Breviari d'Amor* of Matfré Ermengau of Béziers, now in London (BL, Yates Thompson MS 31, fols. 3^v, 131^r, 132^{r-v}, 133^r, 134^{r-v})[53] possibly illuminated in the last quarter of the fourteenth century, the monstrous devils with coloured hybrid bodies with horns, hooves, and tails tie a blindfold over the rabbis' eyes to prevent them from reading if they do not use their diabolic hands to do it; as a result, the rabbis are unable to read their books. Metaphorically, this image could allude to their refusal to believe in the Biblical prophecies of the Old Testament.[54]

But sometimes, the blindfold appears in the scene of the Crucifixion too where the thieves with hooked arms are depicted as blindfolded men. In the historiated initial in the Du Bois Hours in New York (PML, MS M.700, fol. 37^r)[55], for example, that was probably illuminated in Oxford *c*.1325–1330, the thieves are both blindfolded while Christ is not. Again, their lack of sight is not permanent and can be cured by Christ Himself who can redeem it with His death. Even if in the Middle Ages the criminals condemned to death were commonly hooded or blindfolded so as not to be aware of their execution, one may find a connection between the blindfold on the eyes of the thieves and their spiritual blindness both

---

the Margins of Gothic Manuscripts (Berkeley-Los Angeles: University of California Press, 1966), p. 35.

[50] The emphasis on the Jews as the enemies of Christ was prevalent after the Crusades when anti-semitism became increasingly institutionalised in Europe. See Michael Frassetto, *Christian Attitudes towards the Jews in the Middle Ages* (New York-London: Routledge, 2007).

[51] Daniel Boyarin, *Sparks of the Logos: Essays in Rabbinic Hermeneutics* (Leiden: Koninklijke Brill NV, 2003), p. 197.

[52] Barasch, p. 85.

[53] See the *Catalogue of Illuminated Manuscripts* <http://prodigi.bl.uk/illcat/record.asp?MSID=8117&CollID=58&NStart=31> [accessed 16 January 2013].

[54] See Boyarin, *Sparks of the Logos*, p. 55.

[55] See *Corsair* <http://utu.morganlibrary.org/medren/single_image2.cfm?imagename=m700.037ra.jpg&page=ICA000108419> [accessed 16 January 2013].

to Christ's divinity and to the significance of His sacrifice.[56] However, in other cases, Christ Himself is depicted as a blindfolded man, as in the historiated initial at the beginning of the Prime in a Psalter decorated in Liège at the end of the thirteenth century and now in New York (PML, MS M.183, fol. 222^r).[57] Christ is shown here before Pilate with the eyes covered. In the scene of the Mocking too, Christ is blindfolded in three full-page English and French miniatures in De Lisle Hours (New York, PML, MS G.50, fol. 57^v)[58] illuminated in York between 1316 and 1331, in a Book of Hours in New York (PML, MS M.90, fol. 47^r)[59] produced perhaps in Verdun and Paris in 1375, and in the Fitzwarin Psalter in Paris (Bibliothèque nationale, MS lat. 765, fol. 10^r)[60] illuminated in the second half of the fourteenth century.[61] Depicted as the seated Saviour, Christ is suffering mockery respectively from two, three, and four torturers who continue incessantly to inflict pain, suffering, and humiliation on Him. Christ accepts the human form of his torment and endures unbearable pain in solitude to redeem the world. He knows what is going to happen in His immediate future and is represented as a sacrificial victim who controls His destiny; probably for this reason, the blindfold is transparent in the second psalter mentioned.

Even in medieval games and courtly pastimes illuminated in the margins of some sacred and secular codices, the inability to see is parodied and used for the amusement of the onlookers. Blind people are shown as the objects of injury and mortification, and consequently of ridicule. In the famous Flemish copy of the *Romance of Alexander the Great* completed in Tournai in 1344 — with some English additions datable to the beginning of the fifteenth century — and now in Oxford

---

[56] See Mitchell B. Merback, *The Thief, the Cross and the Wheel: Pain and Spectacle of Punishment in Medieval and Renaissance Europe* (London: Reaktion Books, 1999), pp. 22, 61, 221.

[57] See *Corsair* <http://utu.morganlibrary.org/medren/single_image2.cfm?imagename=m183.222ra.jpg&page=ICA000114684> [accessed 16 January 2013].

[58] See *Corsair* <http://utu.morganlibrary.org/medren/single_image2.cfm?imagename=g50.057v.jpg&page=ICA000001579> [accessed 16 January 2013].

[59] See *Corsair* <http://utu.morganlibrary.org/medren/single_image2.cfm?page=ICA000117050&imagename=m90.047r.jpg> [accessed 16 January 2013].

[60] See *BnF Banque d'images - Picture collection* <http://visualiseur.bnf.fr/Visualiseur?Destination=Daguerre&O=8011384&E=JPEG&NavigationSimplifiee=ok&typeFonds=noir> [accessed 16 January 2013].

[61] Freeman Sandler, 133–135.

(Bodleian Library, Bodley MS 264),[62] a continuous series of illuminations occupies the lower margins and forms a parallel narration interacting with the historiated initial.[63] A strange and rather repulsive game is depicted in the bas-de-page on fol. 74^v (fig. 4) where the miniatures opening this section of the book depicts Alexander killing his favourite bleeding horse Bucephalus whose shins were fatally wounded by Pyrrhus. On the left side of the margin of the page, a group of four blind men carrying clubs is led by a boy, while on the right, one of the four men knocks his falling companion. The scene evokes the text on the *Garçon et l'Aveugle*,[64] but is also possibly reminiscent of the biblical image of the blind leading the blind mentioned in both the Gospels of Matthew (15. 14) and Luke (6. 39). However, in the conventional iconography, several blind beggars 'each holding on to the next for guidance, and having a stick in his other hand'[65] are depicted in the countryside and the leader is commonly shown tumbling into the ditch with the others following him. Recently, Edward Wheatley has examined the marginal illumination of the *Romance of Alexander* interpreting it as a 'cruel, humiliating, and potentially deadly "game"'[66] in which blind men were obliged to hurt themselves only to amuse the spectators. In 2005, Wheatley has connected this picture to a spectacle which was observed in Paris in 1425 and narrated by an anonymous chronicler. His translation reads:

> *Item*, on the last Sunday of the month of August there took place an amusement at the residence called d'Arminac in the rue St Honoré, in which four blind people, all armed, each with a stick, were put in a pen, and in that location there was a strong pig that they were to have if they could kill it. Thus it was done, and there was a very strange battle, because they gave themselves so many great blows with those sticks that it went worse for

---

[62] The digital reproductions of the whole manuscript are on the website *Early Medieval Manuscripts at Oxford University* <http://image.ox.ac.uk/show?collection=bodleian&manuscript=msbodl264> [accessed 16 January 2013].

[63] On the manuscript, see for example the recent publication of Mark Cruse, *Illuminating the Roman d'Alexandre, Oxford, Bodleian Library, MS Bodley 264: The Manuscript as Monument* (Cambridge: D.S. Brewer, 2011).

[64] Philippe Ménard, 'Les illustrations marginales du "Roman d'Alexandre"' (Oxford, Bodleian Library, Bodley 264), in *Risus Mediaevalis: Laughter in Medieval Literature and Art* (Leuven: University Press, 2003), pp. 75–118 (p. 84).

[65] James Hall, *Dictionary of Subjects and Symbols in Art* (London: John Murray Publishers, 1974), p. 49.

[66] Edward Wheatley, 'The Blind Beating the Blind: An Unidentified "Game" in a Marginal Illustration of the Romance of Alexander, MS Bodley 264', *Journal of the Warburg and Courtauld Institutes*, 68 (2005), 213–217 (p. 215).

them, because when the stronger ones believed that they hit the pig, they hit one another, and if they had really been armed, they would have killed one another. *Item*, the Saturday evening before the aforementioned Sunday, the said blind people were led through Paris all armed, with a large banner in front, where there was a pig portrayed, and in front of them a man playing a bass drum.[67]

Although this text was written in France eighty years after the completion of the *Romance of Alexander*, the description of the spectacle corresponds to the illumination for the number of the protagonists, the presence of the leading boy and the pig. This animal is substituted with a sow in the manuscript probably because of its reputed fierceness, Wheatley suggests.[68] An established set of rules must have directed this kind of competition repeatedly performed in Europe in the fifteenth century but also before. The aim of the game was evidently to receive the pig – or the sow in this case – as a prize. But the blind players are treated as animals led by a boy with a rope to a pen; one can ask if their blindness and physical imperfection made them look so disgusting and unpleasant as to resemble animals to medieval onlookers. In addition, we can presume that the spectators were less interested in the winner of the pig than in seeing people injure each other. The troubling phenomenon is that people beheld what other people suffered, as if the corporal violations inflicted on blind people could entertain and amuse the other onlookers.[69] This illumination mirrors human cruelty that was familiar and tolerable in the Middle Ages. But in all probability, this game was not conceived of as an image of atrocity distorted into an instrument of pleasure, as we can see it nowadays.

Playful and lively games involving blindfolded and hooded people are certainly less atrocious, but they presuppose in any case that the momentary blind player is in an unfavourable condition. He is under the other players' control, but the difference from the already considered brutal activities mainly consists of the temporary blindness that is part of the rules and a fundamental prerequisite for the game. So, the central player assumes a very important function for the enjoyments of both onlookers and participants. Again, the *Romance of Alexander* in Oxford offers some interesting hints among its 186 marginal illustrations.[70] In

[67] Ibid.
[68] Ibid., p. 216.
[69] See Merback, pp. 7–9, 125.
[70] Ménard, 'Les illustrations marginales', p. 76.

some plays depicted on the foliate bars of the lower borders, a hooded player is surrounded by other players who get pleasure from hitting him with a knotted cloth as for example on fols. 130^{r-v} where the player is waiting to be struck with his legs apart, probably trying to offer resistance to the expected moment of the attack.[71] The other players are tying a knot in the cloth to plan a good attacking play, if they are not about to hit the main player.

Another very popular pastime is called in English 'hot cockles', but this term refers to two distinct games known in French as *hautes coquilles* and *la main chaude*.[72] In the first, the player in the middle hides his head in somebody's lap and becomes momentarily blind in this way; 'a buffet is given by one of the group, who must be identified or the process is repeated'. In the second, the head of the player is hidden again, but he places his hand behind his back and 'the hand is struck by one of the party'[73] and he has to identify his assailant. Both these versions appear on ivory writing tablets where the players usually crowd around the 'blind' man, raising their hands to hit him.[74] In some cases, the players devote themselves to other amusements during the game: they make conversation, embrace, or kiss, and such diversions evidently distract them from the play. In the bas-de-page of a Parisian Breviary intended for Franciscan use and now in the University Library in Cambridge (University Library, MS Dd. 5. 5, fol. 280^r) illuminated *c.*1330–1340,[75] for example, only one man is hitting the central player who is hiding his head in a lady's lap, while two other men are cooking food in a wood-burning oven on the other side of the decorative foliate bar.[76]

---

[71] See *Early Medieval Manuscripts at Oxford University* <http://image.ox.ac.uk/show?collection=bodleian&manuscript=msbodl264> [accessed 16 January 2013].

[72] Richard H. Randall, 'Frog in the Middle', *The Metropolitan Museum of Art Bulletin*, 16 (1958), 269–275 (p. 270).

[73] Ibid.

[74] See, for example, the ivories published on the website *Gothic Ivories* of The Courtauld Institute of Art <http://www.gothicivories.courtauld.ac.uk/search/results.html?qs=hot+cockles+> [accessed 16 January 2013].

[75] Paul Binski and Patrick Zutschi, *Western Illluminated Manuscripts: A Catalogue of the Collection in Cambridge University Library* (Cambridge: University Press, 2011), pp. 299–300.

[76] Randall, p. 30. In this illuminated examplar, as in other ivories with a courtly background, the game might have some sexual connotations. See Raymond Koechlin, *Les ivoires gothiques français*, 3 vols (Paris: F. De Nobele, 1968), II, 424.

In a similar game called 'the frog in the middle' (in French *grenouille*) recurring on many ivories[77] and manuscript marginalia, as in famous French *Chansonnier de Paris* (Montpellier, Bibliothèque de la Faculté de Médicine, MS H.196, fol. 88ʳ), the player representing the frog is covering his eyes with his right hand, 'apparently counting with his eyes closed at the beginning of the game'.[78] The player is seated on the ground with his legs crossed while the others form a ring gambolling about him, buffeting him, or pulling his hair; to escape from this predicament, he has to catch one of them without uncrossing his legs or rising from the floor but he cannot see them moving around. This is the only example detected where the frog in the middle is blind and has to identify those who are playfully dancing around, buffeting him, pushing him, and pulling his hair. By contrast, in the already mentioned *Romance of Alexander* (MS Bodley 264, fols 65ʳ, 97ᵛ, 130ʳ⁻ᵛ) or in the famous Hours of Jeanne d'Evreux (New York, The Metropolitan Museum of Art, The Cloisters, Acc. 54.1.2)[79] illuminated in Paris *c.*1324–28, the frog in the middle has his eyes open. In these illuminations, the players usually dance about and gesture towards the central player; they move around with exaggerated movements, 'executing various capers and at times appearing like a *corps de ballet*'.[80] In these games, blindness seems not to be a sign of depravity and sin. It is simply ridiculed and parodied.

To conclude, this selection of manuscripts contributes to document the multiplicity of connotations that blindness has in the Middle Ages. Although the full meaning of these images is only partially accessible to us now, these images offer an exemplary mirror of medieval culture and thought. They do not simply translate the textual contents into concrete pictures condensing the message into comprehensible visual examples that the medieval viewer could understand more easily, but they are keys to our interpretation of the medieval period. They provide us with fundamental information on the way in which blindness and blind people were considered both in the religious and secular contexts. As either

---

[77] For example, see the website *Gothic Ivories* <http://www.gothicivories.courtauld.ac.uk/search/results.html?ixsid=6WR7V1gr6Nl&qs=frog+in+the+middle> [accessed 16 January 2013].

[78] See Randall, 272. For this manuscript, see also François Avril, 'Chansonnier de Montpellier', in *L'art au temps des rois maudits*, pp. 262–264, at least.

[79] The image is on the website of the Metropolitan Museum <http://www.metmuseum.org/collections/search-the-collections/70010733?img=7> [accessed 16 January 2013].

permanent, or momentary disease caused by God or by a blindfold, blindness was an emblem of the lack of Christian faith and a symbol of sin. It was a sign of the punishment given by God to wicked people. Sightless people appear responsible for their handicap and were thus condemned for the impious conduct that caused their blindness. Blindness was thus thought to express their inner immorality. They were consequently discriminated against because of their condition and exiled as social outcasts who were rejected by contemporary society. In general, blindness was regarded as a good reason for exclusion and humiliation. In the lay context of private and public games and pastimes, blind people as either blind from birth or blindfolded were mocked and parodied. Even in those plays where blindness is temporary and caused simply by a blindfold or a hood, sightless people are humiliated and subjected to mockery and derision like Christ in the Passion. Again, blindness is connected with mistreatment and suffering. Blindness would seem to be a question of inferiority in any case and nobody would take the place of a blind man unless his assailant were identified to take the victim's place.[81]

---

[81] I would like to thank sincerely Ennio Ferraglio, Director of the Biblioteca Queriniana in Brescia, Mons. Pierantonio Gios, Director of the Biblioteca Capitolare in Padua, the British Library and the Bodleian Library (particularly, Tricia Buckingham and Anne Mouron) for the permission to use the copyrighted materials, and Alessandra Petrina for her constant support.

Zonno, Fig. 1. Brescia, Biblioteca Queriniana, MS A. V. 17, fol. 7ʳ.

Zonno, Fig. 2. London, British Library, MS Yates Thompson 13, fol. 103r.

Zonno, Fig. 3. Padova, Biblioteca Capitolare, MS B. 14, fol. 177ʳ.

Zonno, Fig. 4. Oxford, Bodleian Library, MS Bodley 264, fol. 74v.

LIGHT AND VIRTUE:
THE GLOUCESTER CANDLESTICK

Stephanie Seavers and Catia Viegas Wesolowska

The Gloucester Candlestick (fig. 1), in the Victoria and Albert Museum, is a significant and rare survival of Romanesque metalwork. Its exquisite and complex decoration demonstrates the expert artistry of twelfth-century craftsmen. Entwined hybrids and beasts clamber endlessly across its form, struggling towards the light of the candle or tumbling into the shadow below. The secure dating and intriguing provenance of the candlestick further elevates the historical significance of the piece. An inscription upon the stem reveals that Abbot Peter gave the candlestick to the Church of St Peter in Gloucester (now Gloucester Cathedral),[1] dating it between 1104 and 1113, the years of his rule as abbot. Another inscription, believed to be medieval and situated on the inside rim of the drip pan notes that Thomas Pociensis (an unidentified figure) gave the candlestick to Le Mans Cathedral, where it remained until the eighteenth century.[2] The craftsmanship and provenance of the candlestick has placed it at the centre of studies of medieval art and culture. Yet its symbolic role in expressing the beliefs and ideals of the medieval church and its community has been somewhat overlooked. Recent research undertaken at the Victoria and Albert Museum has sought to consider this symbolism. Art historical investigation reveals that the Gloucester Candlestick was valued not only as a bearer of light, but also as a *symbol* of light, which reflected the varied meanings of light and shadow inherent in medieval culture. Conservational analysis meanwhile provides evidence that the strength of this symbolism pervaded the very design and construction of the object.

---

[1] ABBATIS PETRI GREGIS ET DEVOTIO MITIS ME DEDIT ECCLESIE SCI PETRI GLOECESTRE (Abbot Peter's Flock and [his] humble devotion gave me to St Peter's Church in Gloucester).

[2] HOC CENOMANNENSIS RES ECCLESIE POCIENSIS/THOMAS DITAVIT CVM SOL ANNVM RENOVAVIT (Thomas of Poché gave this to the Church of Le Mans, when the sun renewed the year).

Previous research on the Gloucester Candlestick has focused on its origins and decorative style. Though made for an English church, the decoration of the candlestick can be compared to continental examples, in particular two eleventh century candlesticks of electrum (an alloy of gold and silver) from Hildesheim in Germany, which share the Gloucester Candlestick's three knop form, tripod base and decoration of beasts and figures.[3] Iconographic studies of the candlestick however have revealed close similarities with English manuscript illumination. Charles Oman and Anabell Harris both considered the candlestick to be English, relating its decoration to artistic compositions in manuscripts from Durham and Canterbury respectively.[4] Harris particularly noted that the varied and unrepeated detail of the candlestick is in the spirit of English Romanesque art as seen in manuscript illumination. This 'inventiveness' is distinct from the more static geometric forms of German art demonstrated on the Hildesheim candlesticks.[5] More recently Malcolm Thurlby has argued for the English origins of the candlestick and noted its associations with the style of the so-called Hereford school, whose Romanesque architecture is evident in the surrounding areas of Gloucester.[6] The architecture of the Church at Kilpeck, Herefordshire, shows a resemblance to the candlestick, which may indicate a local artistic style.[7] Only Alan Borg's study of the candlestick has sought to look beyond issues of provenance to consider the use and meaning of the piece. He has suggested that the candlestick was placed upon the altar, perhaps to illuminate a shrine or relics, which were put upon the altar from around the year 1000. There is no documentary evidence to suggest that candesticks were used on the altar in this period. However, an eleventh-century ivory from the V&A appears to depict a candlestick upon an altar, providing visual evidence for this usage at an early date.[8]

---

[3] Alan Borg, 'The Gloucester Candlestick', in *Medieval Art and Architecture at Gloucester and Tewksbury, British Archaeological Association,* 7 (1985), 84–92 (p. 88).

[4] Charles Oman, 'The Gloucester Candlestick', in *Victoria and Albert Museum Monographs,* 11 (1958), 1–14 and Anabell Harris, 'A Romanesque Candlestick in London', *Journal of the British Archaeological Association,* 27 (1964), 32–62.

[5] Harris, 'A Romanesque Candlestick', p. 40.

[6] Malcolm Thurlby, *The Herefordshire School of Romanesque Sculpture* (Hereford: Logaston Press, 1999), pp.18–19.

[7] Thurlby, *The Herefordshire School,* p. 18.

[8] Borg, 'The Gloucester Candlestick', pp. 86–88.

Borg also considered the significance of the Gloucester Candlestick within the medieval church. A third inscription around the outside of the drip pan refers to the light of virtue and the shadow of sin:

> LUCIS ON(US) VIRTUTIS OPUS DOCTRINA REFULGENS PREDICAT UT VICIO NON TENEBRETUR HOMO
>
> [Light's burden, virtue's work, brilliantly shining teaching preaches so that Man may not be darkened by sin.]

Through his translation, Borg suggests that the light of the candle symbolises Christ.[9] The decorative figures rendered beneath represent 'a confusion of evil'; a metaphor for the battle against sin and the struggle to reach the light of God.[10] These considerations provide the basis for further study. Comparative analysis of the inscription with evidence of the symbolism of light in medieval culture indicates that the symbolic meaning of the candlestick is more complex than Borg suggests. The significance of the piece had many layers, symbolised not only through its usage but also its decoration, materials and form.

The inscription on the drip pan demonstrates a clear connection between the candlestick and the symbolism of light. The words *lucis*, *virtutis*, *vicio* and *tenebretur* indicate a symbolic balance which connects light with virtue and shadow with vice. Yet beyond this basic interpretation, the exact meaning of the inscription is unclear. Attempts to translate the words have resulted in many different interpretations, each defining a slightly different sense and symbolism. At the sale of the candlestick from the collection of Monsieur d'Espaulart of Le Mans in 1857, for example, the following translation was published in the catalogue:

> Masse de lumière, oeuvre de vertu, doctrine brillante (ce flambeau) prêche pour que l'homme ne se laisse pas envelopper dans les ténèbres du vice.[11]

One might offer the following literal translation:

> mass of light, work of virtue, shining doctrine (the candlestick) preaches so that man does not allow himself to be enveloped in the shadows of vice.

---

[9] 'This burden of light is the work of virtue, shining doctrine teaches that man be not shadowed by vice' (Borg, 'The Gloucester Candlestick', p. 84).

[10] Borg, 'The Gloucester Candlestick', p. 89.

[11] *Catalogue de la Précieuse collection d'objets d'art et Haute Curiosité de Monsieur d'Espaulart du Mans*, 7–9 Mai Hôtel des Ventes Mobilières, Rue Druot (Paris, 1857).

This translation places the candlestick at the centre of light symbolism. The light of the candle symbolises virtue, while the candlestick itself represents shining doctrine. As such, the candlestick is at the heart of the action of the words, the very symbol of the preaching of the word of God. Just as the candlestick is the medium through which the light (of virtue) is beheld, so doctrine is the medium which allows the sinner to free himself from the darkness of vice. Such an understanding of the inscription suggests a profound significance for the candlestick, indicating that it functioned not only as a practical object within the medieval church, but as a spiritual medium for the light of God. Yet the translation relies on the assumption that *flambeau*, omitted from the inscription, signifies *doctrine brillante*. While *onus* (burden) may allude to the candlestick in this sense, the Latin makes no direct reference to it. Thus Charles Oman in 1958 was able to determine a wholly different meaning:

> The debt of light is the practice of virtue. The glorious teaching of the Gospel preaches that man be not benighted in vice.[12]

Here, the significance of the candlestick is eliminated from the focus of the inscription's meaning. The translation of *onus* as 'debt', rather than 'burden' further detracts attention from the function of the candlestick as a symbol. Any other allusion to light has also been removed. *Refulgens* becomes 'glorious' rather than its more literal meaning 'reflecting' or 'glistening'. In fact, this rather enigmatic translation displays only a basic symbolic meaning: that light is a symbol of virtue and shadow is a signification of vice.

Another translation of the inscription by Neil Stratford in 1984 highlighted a more complex association with the symbolism of light, yet similarly overlooked any possible connection to the candlestick:

> This flood of light, this work of virtue, bright with holy doctrine instructs us, so that man shall not be benighted in vice.[13]

According to this translation, there is a multi-layered symbolism to the inscription. Light is associated with the candle, with virtue and with doctrine. Yet the candlestick is only the bearer of symbolic light, not a symbol itself. Colin Sydenham however, in his discussion of the

---

[12] Oman, 'The Gloucester Candlestick', p. 1.
[13] *English Romanesque Art 1066–1200*, ed. George Zarnecki, Janet Holt and Tristram Holland (London: Arts Council of Great Britain in association with Weidenfeld and Nicholson, 1984), p. 249.

inscriptions, elaborated these varied layers of meaning.[14] Interpreting the inscription as an elegiac couplet, he suggested that the rhythm of the words enables the identification of their meaning. According to this approach, *Doctrina* is nominative (with a short a) and is the subject of *praedicat*. The first four words of the inscription stand out as a separate part, governed by the implied yet omitted verb, *est*. Translating the words basically as 'The burden of light is the work of virtue. Doctrine shines forth and preaches so that man is not overshadowed by vice', Sydenham deduces that shadow symbolises vice, the shining light of the candle represents doctrine and the bearer of this light, the candlestick, signifies virtue. He interpreted this symbolism in both prose and verse translations:

> Carrying the candle is the task of righteousness: its light is the Church's teaching, whose message redeems man from the darkness of vice.

And:

> Virtue bears the candle:
> God's word is the flame.
> Man's rescue from sin's shadow
> Together they proclaim.[15]

While Sydenham's interpretation does not take into account the evident balance and even spacing between *lucis*, *virtutis*, *vicio* and *tenebretur* in the inscription, the multi-layered symbolism he suggests is intriguing. It provides a context in which the candlestick itself could be viewed as a symbol and demonstrates the varied messages it may have conveyed. It could of course be argued that the inscription to some extent is insignificant owing to the fact that few people would have seen or read it. Yet a study of the meaning of light in religious and popular culture demonstrates that the symbolisms described in this inscription were simply reinforcing meanings already inherent in medieval society. For those unable to read the words, the function of the object and its use in church would have instilled its symbolic meaning. Further evidence suggests that the very materials, form and decoration of the candlestick also sought to convey the symbolism of light to its audience.

The use of lighting in the early church took a primarily functional role. Placed around the church, candlesticks provided light for both priest and

---

[14] Colin Sydenham, 'Translating the Gloucester Candlestick', in *The Burlington Magazine*, 126.977 (1984), 504.

[15] Sydenham, 'Translating the Gloucester Candlestick', p. 504.

congregation in hours of darkness, particularly in areas around tombs or the shrines of relics.[16] By the turn of the twelfth century however, lighting had taken on symbolic meaning. Documentary evidence shows that large candlesticks were commonly donated to English churches as a sign of faith and benevolence. For example King Cnut is documented to have given a silver seven-branched candlestick to Winchester in 1035.[17] The two large Benedictine houses at Canterbury also record the donation of large candlesticks. Conrad, prior of Christchurch, between 1107 and 1126, gave a 'candlestick of remarkable size made of brass, with three branches on either side springing from the main stem, on which seven candles were placed'.[18] Similarly Abbot Hugh de Fion (1091–1124) presented a candlestick said to be 'from foreign parts' to St Augustine's Abbey.[19] Abbot Peter's gift to the Benedictine Church at Gloucester appears to be part of this trend, in which benefactors marked their importance through the donation of a significant gift. The small size of the Gloucester Candlestick in comparison to these many-branched freestanding holders may not have held the same extravagance, yet Abbot Peter's more intimate gift was similarly significant, possibly even a gift to mark his benediction in 1107. The inscription suggests that the meaning and purpose of the object was of importance to Abbot Peter and his flock. At about 51cm tall, the candlestick is perhaps too large to be one of a pair of candlesticks for a shrine or tomb, but too small for a freestanding piece. Yet the inscription implies through the words *doctrina* and *praedicat* that the candlestick participated in some way in the prayers of the Christian church. Could the candlestick have been used to illuminate the bible, or another Christian text, thus allowing the doctrine of God to 'shine forth'? Was this candlestick, as some translations of the inscription suggest, a medium for the light of God, providing light with which the priest could preach His word? Alternatively the candlestick may have been used in Christian ceremony, visually symbolising the sentiments displayed on its inscription through its use. Within the large Benedictine community of Gloucester, there were many opportunities for such a beautiful object to play a significant role. Light was an important aspect of Benedictine life. The prologue to the Rule of St Benedict calls for believers to 'open our eyes to the light that comes from God' (*apertis*

---

[16] Borg, 'The Gloucester Candlestick', p. 86.
[17] D. R. Dendy, *The Use of Lights in Christian Worship* (London: SPCK, 1959), p. 14.
[18] Dendy, *Lights in Christian Worship*, p. 13.
[19] Oman, 'The Gloucester Candlestick', p. 10.

oculis notris ad deificum lumen). The passage uses the image of awaking from sleep as an allegory for the realisation of the divine:

> Exurgamus ergo tandem aliquando excitante nos scriptura ac dicente: *Hora est iam nos de somno surgere*, et apertis oculis nostris ad deificum lumen attonitis auribus audiamus divina cotidie clamans quid nos admonet vox dicens: *Hodie si vocem eius audieritis, nolite obdurare corda vestra*. Et iterum: *Qui habet aures audiendi audiat, quid spiritus dicat ecclesiis*. Et quid dicit? *Venite, filii, audite me; timorem Domini docebo vos. Currite dum lumen vitæ habetis, ne tenebræ* mortis *vos conprehendant*.
>
> [Let us get up then, at long last, for the Scriptures rouse us when they say: *It is high time for us to arise from sleep* (Romans 13. 11). Let us open our eyes to the light that comes from God, and our ears to the voice from heaven that every day calls out this charge: *If you hear his voice today, do not harden your hearts* (Psalm 94/95. 8). And again: *You have the ears to hear, listen to what the Spirit says to the churches* (Revelations 2. 7). And what does he say? *Come and listen to me, sons, I will teach you the fear of the Lord* (Psalm 33/34. 12). *Run while you have the light of life, that the darkness* of death *may not over take you* (John 12. 35).][20]

The use of light and sound to know God, as expressed in this passage, is echoed in the function and symbolism of the Gloucester Candlestick. As Benedict calls 'let us open our eyes to the light that comes from God and our ears to the voice from heaven', the Gloucester Candlestick responds by providing this light (the 'work of virtue') and announcing the 'shining teaching' of God's word. If the Gloucester Candlestick was indeed used to light religious texts so that the word of God could be spoken and heard, it would have directly represented the theme of enlightenment through sound and sight indicated in the Benedictine Rule.

Contrasts of light and shadow also appear upon the decoration of the Gloucester Candlestick. The struggling figures strive towards the light of the candle or tumble into the shadow below. This movement from darkness to light parallels the light metaphor within the passage, in which the transition from the blindness of sleep to the vision of waking life represents the transition from sinner to believer.

The importance of light in the Benedictine community was not confined to religious allegory. Monastic practices also placed a great significance on light. In their consideration of the medieval 'demonisation' of the night, Deborah Youngs and Simon Harris have

---

[20] *The Rule of St Benedict in Latin and English*, ed. Timothy Fry (Minnesota: Liturgical Press, 1981), pp. 158–9, v. 8.

noted how the Benedictine rule insisted that monks employed their learning 'unceasingly, day and night', thus maintaining the strength of God's benevolence in the community even in hours of darkness. Indeed, Matins, which began in darkness at 2 a.m., were the longest of all the offices.[21] The practice of praying throughout the night to ward off the evil of darkness reflects the sentiment of the Gloucester Candlestick that doctrine may prevent man falling under shadow.

In Church, the use and symbolism of light in certain specific ceremonies reiterate the meaning found upon the Gloucester Candlestick. During the final days of Holy Week, the office of *Tenebrae* was celebrated, in which candles in front of the altar were gradually extinguished to leave only one burning. This ceremony signified Christ's final hours before his sacrifice, when he warned his followers 'walk while ye have the light, lest darkness comes upon you'[22] ('dixit ergo eis Iesus adhuc modicum lumen in vobis est ambulate dum lucem habetis ut non tenebrae vos conprehendant et qui ambulat in tenebris nescit quo vadat', John 12. 35).[23] The Gloucester Candlestick with its decoration and inscription would have heightened the symbolic meaning of such a ceremony. Light was also used at other important points in the year. At Candlemas, which took place on the 2 February, lights were used to symbolise the retreat of darkness from the day. The congregation took candles home with them as spiritual symbols.[24]

Similar evidence suggests that the object continued to be used as a significant ceremonial symbol at Le Mans Cathedral. Cathedral Vespers for example, included ritual lamp lighting and the singing of a hymn of light to mark the close of the day and the retreat of the light.[25] Yet perhaps the most intriguing evidence for the ceremonial use of the Gloucester Candlestick at Le Mans is the inauguration of the cathedral in 1254. To mark the occasion, the congregation took part in an elaborate

---

[21] Deborah Youngs and Simon Harris, 'Demonizing the Night in Medieval Europe: a Temporal Monstrosity?', in *The Monstrous Middle Ages*, ed. Bettina Bildhauer and Robert Mills (Cardiff: University of Wales Press, 2003), pp. 134–154 (p. 141).

[22] Youngs and Harris, 'Demonizing the Night', p. 136.

[23] Citing *Biblia Sacra iuxta Vulgatam versionem*, adiuvantibus Bonifatio Fischer et al., recensuit et brevi apparatu critico instruxit Robertus Weber, ed. quartam emendatam (Stuttgart : Deutsche Bibelgesellschaft, 1994), p. 1682.

[24] Youngs and Harris, 'Demonizing the Night', p. 141.

[25] Michael G. Powell, *Introduction to Medieval Christian Liturgy* <http://www.yale.edu/adhoc/research_resources/liturgy/hours.html> [accessed 16 January 2013].

ceremony of light in which they had candle holders made for the occasion and walked in procession through the Cathedral. According to the *Actes* of Le Mans, the light of the many candles symbolised the faith in the hearts of the congregation.[26] While it is not conclusive that the Gloucester Candlestick was given to Le Mans Cathedral for this occasion, this important ceremony reflects the symbolism of the object entirely. It also provides important evidence that candlesticks were actively used in Christian ceremony and that they held a symbolic significance separate from the symbolic light of the candles they bore. Indeed, the candlesticks in use at the inauguration of Le Mans were of such symbolic importance that according to the *Actes*, certain guilds that had not provided candlesticks recognised their error and proposed to furnish the cathedral with stained glass to light the church to demonstrate their faith.[27]

The varied evidence of the use of light in Christian ceremony during the medieval period demonstrates that candlesticks were used to enact the light of Christian faith and the shadow of sin. The inscription upon the Gloucester Candlestick strongly suggests that it played a highly symbolic role in such ceremonial ritual. The use and significance of the candlestick went far beyond its practical function. It wholly reflected the spiritual symbolism of light. Furthermore, the significant ceremonial use of light and the exquisite decoration and inscriptions upon this object indicate that the candlestick was valued as a symbol in its own right. Candlesticks were lasting symbols of the light of Christ, of virtue, of faith and of truth, even when the spark of the candle had gone out.

The cultural context of light and shadow in the Middle Ages demonstrates the effect of popular belief upon the function, use and symbolism of the objects such as the Gloucester Candlestick. The profound meanings of light and dark defined by medieval people came to be reflected in the objects around them. Yet the symbolism of light also pervaded the very creation of the Gloucester Candlestick, demonstrating that popular cultural ideas influenced the making of medieval objects. Conservational analysis shows that each aspect of the candlestick's construction from the alloy of the metal to its openwork form was carefully considered. As such, the candlestick not only symbolised the light of God through its function and inscription, but also through its material, form and decoration.

---

[26] Michel Bouttier, *La Cathédrale du Mans* (Le Mans: Edition de la Reinette, 2000), p. 77.

[27] Bouttier, *La Cathédrale du Mans*, p. 78.

The material of the Gloucester Candlestick suggests that light played a part in the very essence of the piece. The metal alloy of the candlestick is highly unusual and as yet, no comparative alloy has been found from the middle ages. Analysis carried out in the 1980s, and confirmed more recently, identifies the metal as a brass (an alloy of copper and zinc) with an unusually high content of silver.[28] This addition of silver, an expensive material in this period, is difficult to explain. Previous studies have argued that the alloy was made from the available metals to hand. The addition of silver, perhaps, came from donations of the congregation, and these were melted down to add to the volume of the metal.[29] However there are several problems with this theory. The decoration of the candlestick is a complex openwork form covered in figurative beasts. Although made in three sections, the casting process (using the *cire perdue* or 'lost wax' method) for this object would have been highly complex. It seems unlikely therefore, that an expert craftsman would have added any metal without knowing how it might affect his casting. It is possible that the silver was added to help the flow of the metal through the complex mould. The object certainly has very few flaws despite the detailed nature of its design. However, other cheaper metals could be used for this purpose. The silvery tone to worn parts of the candlestick (fig. 3) suggests another possibility: that the addition of silver was used to make the object look like silver. A thirteenth-century discussion of alchemical theory attributed to Thomas Aquinas argued that copper could be made to look like silver. It advised:

> add to copper some white sublimated arsenic and you will see the copper turn white. If you then add some pure silver you transform the copper into a veritable silver.[30]

The theory that a substance could be changed by adapting the balance of its elements was an idea developed from ancient thinkers such as Aristotle, according to whom all substances consisted of prime matter. Substances differed according to the ways in which their four qualities, heat, cold, dry and moist, were combined. Thus different metals developed from varying interactions between earth, sun and water. This theory continued into the medieval period leading Albertus Magnus to

---

[28] R. Brownsword, E. E. H. Pitt, & J. Wilkin, 'A Technical Note on the Gloucester Candlestick', in *Medieval Art and Architecture at Gloucester and Tewksbury, Journal of the British Archaeological Association*, 7 (1985), 168–170.

[29] Brownsword, 'A Technical Note on the Gloucester Candlestick', pp. 168–170.

[30] Charles Thompson, *Alchemy and Alchemists* (New York: Dover, 2002), pp. 80–81.

suppose that if 'nature could transform sulphur and mercury into metals by the aid of the sun and stars, it seems reasonable that the alchemist should be able to do the same in his vessel'.[31] It is quite possible therefore, that medieval goldsmiths knew of these theories and put them into practice. A casting experiment carried out as part of this research sought to test this theory. A mould was taken of a small section of the candlestick, which was then cast in the lost wax method using silver copper alloy. The results were intriguing. The finished sample (fig. 2) was much whiter than ordinary brass and had a brilliant lustre and shine to the metal, accentuated by its silver colour. The luminous effect of the metal before gilding may even challenge the assumption that the piece was intended to be gilded originally. The brilliance and lustre of metals were certainly praised in the medieval period. Bartholomew Anglicus for example in his *De proprietatibus rerum* noted that electrum, an alloy of silver and gold, shone brighter than gold or silver.[32] It seems significant that electrum with a high silver content was used to make the Hildesheim candlesticks, the closest surviving comparisons to the Gloucester Candlestick.[33] Is it possible that the use of silvery metals was specifically designed to reflect the light and through their colour signify the light and glory of God?

The composition of the metal has some bearing upon the candlestick's inscription. Connections between the candlestick and *doctrina refulgens* (glistening doctrine) may suggest the addition of silver to visually represent this lustre. The idea that light is 'the work of virtue' (*virtutis opus*) however, appears at odds with the colouration of the metal, which had the potential to deceive. The desire to change the colour of the metal may not have been intended to dupe, a falsity that Aquinas warns against,[34] but rather to achieve by eye what could not be provided in essence. Views of alchemy at this early stage in its Western history had not yet determined the boundaries of its truth and falsity. Indeed, in the twelfth century, Jean de Meun in *Le Roman de la Rose* indicated that

---

[31] Lee Patterson, 'Perpetual Motion: Alchemy and the Technology of the Self', in *Temporal Circumstances: Form and History in the Canterbury Tales* (New York: Palgrave, 2006), p. 168.

[32] Robert Steele, *Medieval Lore from Bartholomew Anglicus* (London: Alexander Moring, The De La More Press, 1905), p. 38.

[33] Borg, 'The Gloucester Candlestick', p. 88.

[34] Thompson, *Alchemy and Alchemists*, pp. 81–82.

alchemy, or the colouration of metals, was an honest pursuit, as long as the practitioner himself was true:

> Car d'argent fin finor sont maistre
> Cil qui 'Alchemie sont maistre,
> Et pois et couleur leur ajoustent
> Par choses que gaires ne coustent.
> Et d'or fin pierres precieuses
> Font il cleres et envieuses;
> Et les autres metaus denuent
> De leur formes, si qu'il les muent
> En fin argent par medicines
> Blanches et tresperçanz et fines.
> Mais ce ne feroient cil mie
> Qui oevrent de sophisterie:
> Travaillent tant com il vivront,
> Ja nature n'aconsivront.[35]

The masters of alchemy produce pure gold from pure silver, using things that cost almost nothing to add weight and colour to them; with pure gold they make precious stones, bright and desirable, and they strip other metals of their forms, using potions that are white and penetrating and pure to transform them into pure silver. But such things will never be achieved by those who indulge in trickery: even if they labour all their lives, they will never catch up with Nature.[36]

Thus the creation of the Gloucester Candlestick, made for an honest purpose, may not have been considered any less virtuous for its deceptive appearance. In fact, the addition of silver may have been viewed as a method of *adding* purity and 'virtue' to an otherwise base metal. Later commentaries also show that the colouration of metals was part of the goldsmith's daily work. The sixteenth-century metallurgist Biringuccio apologetically stated that:

> The art of the goldsmith has a close connection with that of the alchemists because it often makes a thing appear what it is not, as is seen in […]

---

[35] Guillaume de Lorris et Jean de Meun, *Le Roman de la Rose*, ed. Armand Strubel (Paris: Librairie générale française, 1992), pp. 846–7, ll.16139–16152.

[36] *The Romance of the Rose*, trans. Frances Horgan (Oxford: Oxford University Press, 1994), p. 249.

heightening the colour of gold, in whitening silver and also in gilding things that really are of silver, brass or copper but appear to be of gold.[37]

That the colouration of metals was part of a goldsmith's trade suggests that the adaptation of metals to create colour and lustre (and therefore light), may have been a standard element in the creation of medieval metalwork: an element which in the case of the Gloucester Candlestick was intended to have a symbolic effect.

While the lustre of the Gloucester Candlestick's metal reflected the light, invoking the 'shining doctrine' of its inscription, its openwork form symbolised the light of God in another way. It not only added depth to the figures entwined upon it but also created the effect of light and shadow. Evidence of the original construction of the candlestick suggests that the openwork decoration was intended to be seen on every section of the object. Small marks at the join of the base and the stem show that the sections were originally fitted together using metal extensions on the base, which fitted tightly into slots inside the stem. The copper tubing now inserted inside the openwork stem appears to be a later (probably medieval) repair. Thus the original effect of the candlestick may have been an entirely openwork form, with only the pricket obscuring the light shining through the object. This construction once again implies the consideration of light at an early stage in the candlestick's creation. The object was clearly designed to manipulate the light around it to create varying degrees of dark and light. Experiments shining soft light through the object show how light from other sources could cast moving shadows and shapes through the openwork (fig. 6). The effect clearly evokes the contrast of light and dark found in the candlestick's inscription and regularly enacted in the ceremonies of the medieval church.

The manipulation of light and the degrees of light and shade represented both physically and symbolically through the Gloucester Candlestick is further echoed in its decoration. The direction of many of the entangled beasts and figures, climbing upwards towards the light and falling down into the shadow has already been noted as a representation of the candlestick's inscription. Yet the finish of the figures, which varies considerably upon each section, may also indicate symbolic meaning. The base of the candlestick is by far the most complex in its casting. Figures and animals strive for position, each one pulling, grabbing or biting the next in a horizontal motion. These figures are extraordinarily detailed,

---

[37] William Eamon, *Science and the Secrets of Nature: Books of Secrets in Medieval and Early Modern Culture* (Princeton, NJ: Princeton University Press, 1994), p. 117.

from the careful working of their limbs to their facial expressions (fig. 4). The stem of the candlestick appears softer in its detail; the figures are more rounded and less well defined (fig. 3). Upon the drip pan, the creatures are almost formless, lacking details of hands and feet (fig. 5). These differences may be explained as the work of three different hands of varying skill. Yet the shifting degrees of figurative form subtly reflect the varying light and shadow upon the candlestick, and echo the stages of man in his various struggles to know God. Significantly, the area upon which one would expect the least light to fall, the underside of the drip pan, depicts the most deformed figures. These creatures lurk, half formed in the shadows, one figure for example has only one eye. Could these shadowy creatures represent the darkness of those in purgatory, from whom light is withheld until, having been slowly cleansed of their impurities, they return to the light?[38] Or do they represent the danger and evil that lurks in the shadows? Could they be, in reference to the candlestick's inscription, personifications of vice? While it is unclear to what degree the decoration of the Gloucester Candlestick should be interpreted, it is certain that the varied creatures upon it were intended to reflect the meaning of light and shadow in Christian thought.

The Gloucester Candlestick has long been recognised as an important historical survival, yet this research demonstrates the significance of the piece in the medieval world. The candlestick was more than a decorative adornment or a rich gift from a faithful benefactor. It signified the light of Christian faith. This symbolism was considered in every aspect of its creation from the way in which the metal caught the light to the shadows cast through the openwork. It is likely that the candlestick actively participated in the highly symbolic rituals of light within the church and may even have become a symbol of faith at the inauguration at Le Mans. While much of the history of this intriguing object will remain in the shadows, its intricate connection with the symbolism light goes some way to explaining the medieval significance of the object and perhaps to indicating the reasons why it survives today.

---

[38] Youngs and Harris, 'Demonizing the Night', pp. 137–38.

The following images are reproduced courtesy of the Victoria and Albert Museum.

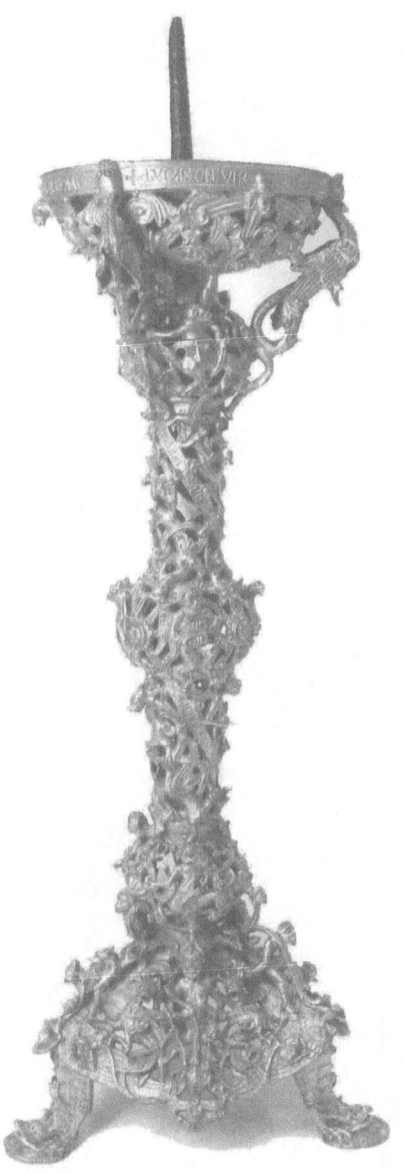

Seavers and Viegas, Fig. 1. The Gloucester Candlestick, Victoria and Albert Museum.

Seavers and Viegas, Fig. 2. Finished sample casting using silver copper alloy.

Seavers and Viegas, Fig. 3. Detail of the Knop.

Seavers and Viegas, Fig. 4. Detail of the base.

Seavers and Viegas, Fig. 5. Detail of beneath the drip pan.

Seavers and Viegas, Fig. 6, Experiment shining light through the openwork.

Abbreviations

CSEL   Corpus Scriptorum Ecclesiasticorum Latinorum.

CCSL   Corpus Christianorum Series Latina.

EETS   Early English Text Society.

PL   Patrologia Latina, ed. J.P. Migne (Paris: 1844-65).

Bibliography

Adelard of Bath, 'Natural Questions', trans. Hermann Gollancz, in *A Sourcebook of Medieval Science*, ed. Edward Grant (Cambridge, MA: Harvard University Press, 1974).

Albertus Magnus, *De vegetabilibus libri VII*, ed. Ernest Meyer (Berlin: G. Reimer, 1867).

Alfayev, H., *St Symeon the New Theologian and Orthodox Tradition* (Oxford: Oxford University Press, 2000).

Alfayev, H., 'The Patristic Background of Symeon the New Theologian's Doctrine of Light', *Studia Patristica*, 32 (1997), 229–238.

Andreas Capellanus, *De Amore (Andreas Capellanus on Love)*, ed. and trans. P. G. Walsh (London: Duckworth, 1982).

Arano, L. Cogliati, ed., *The Medieval Health Handbook: Tacuinum Sanitatis* (New York: Braziller, 1976).

Aristotle, *The Parva naturalia; De sensu et sensibili*, trans. J. I. Beare in *The Works of Aristotle Volume III*, ed. W. D. Ross (Oxford: Clarendon Press, 1908).

Assirelli, M., 'L'immagine dello "stolto" nel salmo 52', in *Il Codice miniato: rapporti tra codice, testo e figurazione, Atti del III Congresso di Storia della Miniatura, Cortona, 20–23 ottobre, 1988*, ed. M. Ceccanti and M. C. Castelli (Firenze: L. S. Olschki, 1992), pp. 19–27.

Augustine, *De Genesi ad litteram libri duodecim*, ed. Joseph Zycha, CSEL, 28, 1 (Vienna: F. Tempsky; Leipzig: G. Freytag, 1894).

Augustine, *De Genesi contra Manichaeos*, ed. Dorothea Weber, Sancti Augustini Opera, CSEL, 91 (Wien: Verlag der Österreichischen Akademie der Wissenschaften, 1998).

Saint Augustine on Genesis: Two Books on Genesis – 'Against the Manichees' and 'On the Literal Interpretation of Genesis: An Unfinished Book', trans. Roland J. Teske, The Fathers of the Church, 84 (Washington, D.C.: Catholic University of America Press, 1991).

Augustine, *Enarrationes in psalmos*, ed. E. Dekkers and J. Fraipont, CCSL, 40 (Turnholt: Brepols, 1956).

Augustine, *In Iohannis euangelium tractacus CXXIV*, ed. Radbod Willems, CCSL, 36 (Turnhout: Brepols, 1954); trans. John W. Rettig, *Tractates on the Gospel of John*, 5 vols (Washington, DC: Catholic University of America Press, 1988).

Avery, M., ed., *The Exultet Rolls of South Italy*, II (all published) Plates (Princeton: University Press, 1936).

Avicenna, *The Canon of Medicine of Avicenna*, trans. O. Cameron Gruner (London: Luzac & Co., 1930).

Avril, F., 'Psautier-livre d'heures de Yolande de Soissons', in *L'art au temps des rois maudits Philippe le Bel et ses fils 1285–1328*, exh. cat. (Paris: Editions de la Réunion des musées nationaux, 1998), pp. 298–300.

Bacon, Roger, *Opus maius*, trans. Robert Belle Burke, 2 vols (Philadelphia: University of Pennsylvania Press, 1928).

Bacon, Roger, *Opus minus* in *Fr. Rogeri Bacon opera quædam hactenus inedita*, ed. J.S. Brewer, Rolls Series (London: Longman, Green, Longman and Roberts, 1859; repr. Nendelin: Kraus, 1965).

Bacon, Roger, *On Signs*, trans. Thomas S. Maloney (Toronto: Pontifical Institute of Mediaeval Studies, 2013).

Baert, B., 'The Healing of the Blind Man at Siloam, Jerusalem: A Contribution to the Relationship Between Holy Places and the Visual Arts in the Middle Ages (Parte I)', *Arte cristiana*, 95 (fasc. 838) (2007), 49–60.

Baert, B., 'The Healing of the Blind Man at Siloam, Jerusalem: A Contribution to the Relationship Between Holy Places and the Visual Arts in the Middle Ages (Parte II)', *Arte cristiana*, 95 (fasc. 839) (2007), 121–130.

Baird, W., 'Visions, Revelation, and Ministry: Reflections on 2 Cor 12: 1–5 and Gal 1: 11–17', *Journal of Biblical Literature*, 104 (1985), 651–62.

Ball, P., *Bright Earth: Art and the Invention of Color* (London: Penguin, 2001).

Banister, Richard, *A Treatise of One Hundred and Thirteen Diseases of the Eyes and Eye-liddes* (London: Imprinted by Felix Kyngston, for Thomas Man, 1622).

Banks, M. Macleod, ed., *An Alphabet of Tales*, 2 vols, EETS, o.s. 126 & 127 (London: Keegan Paul, Trench & Trübner, 1904–1905).

Barasch, M., *Blindness: The History of A Mental Image in Western Thought* (New York: Routledge, 2001).

Barnard, F. Pierrepont, ed., *The Essential Portions of Nicholas Upton's De Studio Militari: Before 1446, trans John Blount c.1500* (Oxford: Clarendon Press, 1931).

Barney, S. A., W. J. Lewis, J. A. Beach et al., trans., *The Etymologies of Isidore of Seville* (Cambridge: Cambridge University Press, 2006).

Saint Basil, Bishop of Caesarea, *The Syriac Version of the Hexaemeron by Basil of Caesarea*, trans. and ed. Robert W. Thomson, CSEL, vols 550–1; Scriptores Syri Tomus 222–3 (Lovanii: Peeters, 1995).

Baur, L., ed., *Die Philosophischen Werke des Robert Grosseteste, Bischofs von Lincoln*, Beiträge zur Geschichte der Philosophie des Mittelalters, 9 (Münster: Aschendorff Verlag, 1912).

Beardsley, M. C., *The Aesthetic Point of View: Selected Essays*, ed. Michael J. Wreen and Donald M. Callen (Ithaca: Cornell University Press, 1982).

Becker, A. H., *Fear of God and the Beginning of Wisdom* (Philadelphia, PA: University of Pennsylvania Press, 2006).

Bellinati, C., 'Antifonario', in *Parole dipinte: la miniatura a Padova dal Medioevo al Settecento*, ed. G. Baldissin Molli, G. Canova Mariani, and F. Toniolo (Modena: Franco Cosimo Panini, 1999), pp. 88–90.

Bennett, A., 'The Transformation of the Gothic Psalter in Thirteenth-Century France, in *The Illuminated Psalter: Studies in the Content, Purpose and Placement of its Images*, ed. F. O. Büttner (Turnhout: Brepols, 2004), pp. 211–221.

Bergen, H., ed., *Lydgate's Fall of Princes*, 4 vols, EETS, e.s. 121–124 (London: Oxford University Press, 1924–1927).

Berlioz, J., et al, *Faire croire: modalités de la diffusion et de la réception des message religieux du IIe au XVe siècle* (Rome: École française de Rome, 1981).

Berlioz, J., et al, *L'aveu, antiquité et moyen-âge: actes de la table ronde* (Paris: École française de Rome, 1986).

Bernard of Clairvaux: Selected Works, trans. Gillian R. Evans (New York: Paulist Press, 2005).

Bernardinello, S., *Catalogo dei codici della Biblioteca Capitolare di Padova*, 2 vols (Padova: Istituto per la Storia Ecclesiastica Padovana, 2007).

Bethurum, D., ed., *The Homilies of Wulfstan* (Oxford: Clarendon Press, 1957).

Bianco, S., 'A Blank Monk in the Rose Garden: Lydgate and the *Dit Amoureux* Tradition', *Chaucer Review*, 34 (1999), 60–68.

Biblia Sacra iuxta Latinam Vulgatam versionem ad codicum fidem, 18 vols (Rome: Typis Polyglottis Vaticanis, 1926–1995).

Biblia Sacra iuxta Vulgatam versionem, adiuvantibus Bonifatio Fischer et al., recensuit et brevi apparatu critico instruxit Robertus Weber, ed. quartam emendatam (Stuttgart: Deutsche Bibelgesellschaft, 1994).

Biernoff, S., *Sight and Embodiment in the Middle Ages* (Basingstoke: Palgrave, 2005).

Binski, P. and P. Zutschi, *Western Illuminated Manuscripts: A Catalogue of the Collection in Cambridge University Library* (Cambridge: Cambridge University Press, 2011).

Bloomfield, M. W., *The Seven Deadly Sins: An Introduction to the History of a Religious Concept, with Special Reference to Medieval English Literature* (Michigan: State University Press, 1952).

Blumreich, K. M., ed., *The Middle English Mirror: An Edition Based on Bodleian Library, MS Holkham misc. 40* (Tempe, AZ: Arizona Center for Medieval and Renaissance Studies, in association with Brepols, 2002).
Boffey, J., ed., *Fifteenth-Century English Dream Visions: An Anthology* (Oxford: Oxford University Press, 2003).
Boffey, J., 'The Reputation of Chaucer's Lyrics in the Fifteenth Century', *Chaucer Review*, 28 (1993), 23–40.
Bonaventure, *Itinerarium mentis ad deum*, ed. and trans. Philotheus Boehner (St Bonaventure, NY: The Franciscan Institute, 1942).
Borg, A., 'The Gloucester Candlestick', in *Medieval Art and Architecture at Gloucester and Tewksbury, British Archaeological Association*, 7 (1985), 84–92.
Bouttier, M., *La Cathédrale du Mans* (Le Mans: Edition de la Reinette, 2000).
Boyarin, D., *Sparks of the Logos: Essay in Rabbinic Hermeneutics* (Leiden: Brill, 2003).
Boyer, C. B., 'Aristotelian References to the Law of Reflection', *Isis*, 36.2 (1946), 92–95.
Brandeis, A., ed., *Jacob's Well*, EETS, o.s. 115 (London: Oxford University Press, 1900).
Brewer, D., 'The Colour Green', in *A Companion to the Gawain-Poet*, ed. Derek Brewer and Jonathan Gibson, Arthurian Studies, 38 (Cambridge: D.S. Brewer, 1997), pp. 181–90.
Bridges, J. H., ed., *The Opus maius of Roger Bacon*, 3 vols (London: Williams and Norgate, 1900).
Bridges, J. H., *The Life and Work of Roger Bacon: An Introduction to the Opus maius* (London: Williams & Norgate, 1914; repr. Merrick, NY: Richwood Pub. Co., 1976).
Brock, S., *The Luminous Eye* (Kalamazoo, MI: Cistercian Publications, 1985).
Brownsword, R., E. E. H. Pitt, and J. Wilkin, 'A Technical Note on the Gloucester Candlestick', in *Medieval Art and Architecture at Gloucester and Tewksbury, Journal of the British Archaeological Association*, 7 (1985), 168–170.
Büttner, F. O., 'Der illuminierte Psalter im Westen', in *The Illuminated Psalter: Studies in the Content, Purpose and Placement of its Images*, ed. F. O. Büttner (Turnhout: Brepols, 2004), pp. 1–106.
Camille, M., 'Before the Gaze: The Internal Senses and Late Medieval Practices of Seeing', in *Visuality Before and Beyond the Renaissance*, ed. R. S. Nelson, pp. 197–223.
Cameron, A., *The Byzantines* (Oxford: Blackwell, 2006).
Campbell, J. J., ed., *The Advent Lyrics of the Exeter Book* (Princeton: Princeton University Press, 1959).

Carlson, D. R., 'The Chronology of Lydgate's Chaucer References', *Chaucer Review*, 38 (2004), 246–54.

Catalogue de la Précieuse collection d'objets d'art et Haute Curiosité de Monseiur d'Espaulart du Mans, 7-9 Mai Hôtel des Ventes Mobilières, Rue Druot (Paris, 1857).

Catto, J. I. and R. Evans, eds, *The History of the University of Oxford, Volume II: Late Medieval Oxford* (Oxford: Oxford University Press, 1992).

Chaucer, Geoffrey, *The Riverside Chaucer*, ed. Larry D. Benson (Oxford: Oxford University Press, 1988).

Chazelle, C., 'Violence and the Virtuous Ruler in the Utrecht Psalter, in *The Illuminated Psalter: Studies in the Content, Purpose and Placement of its Images*, ed. F. O. Büttner (Turnhout: Brepols, 2004), pp. 337–48.

Chidester, D., *Word and Light: Seeing, Hearing, and Religious Discourse* (Chicago: University of Illinois Press, 1992).

Chidester, D., 'The Symbolism of Learning in Augustine', *The Harvard Theological Review*, 76 (1983), 73–90.

Churchland, P. S., *Neurophilosophy: Toward a Unified Science of the Mind-Brain* (Cambridge, MA; London: MIT Press, 1986).

Clegg, B., *The First Scientist: A Life of Roger Bacon* (New York: Carroll & Graf, 2003).

Colish, M. L., *The Mirror of Language: A Study in the Medieval Theory of Knowledge* (New Haven, MA: Yale University Press, 1968).

Collins, M., *Medieval Herbals: The Illustrative Traditions* (London: British Library, 2000).

Corson, R., *Fashions in Eyeglasses* (London: Peter Owen, 1967).

Cottino-Jones, M., 'Fabula vs. Figura: Another Interpretation of the Griselda Story', *Italica*, 50 (1973), 38–52.

Crawford, A., ed., *The Household Books of John Howard, Duke of Norfolk, 1462–1471, 1481–1483*, 2 vols (Stroud: Sutton for Richard III and Yorkist History Trust, 1992).

Crawford, S. J., 'The Worcester Marks and Glosses of the Old English Manuscripts in the Bodleian Together with the Worcester Version of the Nicene Creed', *Anglia*, 52 (1928), 1–25.

Crowley, T., *Roger Bacon: The Problem of the Soul in his Philosophical Commentaries* (Louvain-Dublin: Éditions de l'Institut supérieur de philosophie, 1950).

Cruse, M., *Illuminating the Roman d'Alexandre, Oxford, Bodleian Library, MS Bodley 264: The Manuscript as Monument* (Cambridge: D. S. Brewer, 2011).

Daniele, I., 'Daniele di Padova', in *Bibliotheca Sanctorum*, 13 vols (Rome: Istituto Giovanni XXIII della Pontificia Università Lateranense, 1964), IV, 474–475.

Davidoff, J. M., 'The Audience Illuminated, or New Light Shed on the Dream Frame of Lydgate's *Temple of Glas*', *Studies in the Age of Chaucer*, 5 (1983), 103–25.

Daly, W. J. and R. D. Yee, 'The Eye Book of Master Peter of Spain: A Glimpse of Diagnosis and Treatment of Eye Disease in the Middle Ages', *Documenta Ophthalmologica*, 103 (2001), 119–53.

Dawson, W. R., ed., *A Leechbook or Collection of Medical Recipes of the Fifteenth Century* (London: Macmillan, 1934).

Dendy, D. R., *The Use of Lights in Christian Worship* (London: SPCK, 1959).

Pseudo-Dionysius: The Complete Works, trans. Colm Luibheid, forward, notes, introductions by Paul Rorem, Rene Roques, Jaroslav Pelikan et al., Classics of Western Spirituality (New York: Paulist Press, 1987)

Doane, A. N., ed., *The Saxon Genesis* (Madison, Wisconsin: University of Wisconsin Press, 1991).

Downing, J. Y., 'A Critical Edition of Cambridge University MS Ff.5.48' (unpublished Ph.D. thesis, University of Washington, 1969).

Dronke, P., ed., *Medieval Latin and the Rise of European Love-Lyric*, 2 vols (Oxford: Clarendon Press, 1968).

Duby G., *Saint Bernard et l'art cistercien*, Arts et Métiers Graphiques (Paris: Flammarion, 1979).

Duffy, E., *The Stripping of the Altars : Traditional Religion in England, c.1400–c.1580*, 2nd edn (New Haven, MA: Yale University Press, 2005).

Duncan, T. G., and M. Connolly, *The Middle English Mirror: Sermons from Advent to Sexagesima*, Middle English Texts, 35 (Heidelberg : C. Winter, 2003).

Dutton, P. E., 'Medieval Approaches to Calcidius' in *Plato's Timaeus as Cultural Icon*, ed. G. Reydams-Schils (Notre Dame, IND: Notre Dame University Press, 2003), pp. 183–205.

Eamon, W., *Science and the Secrets of Nature: Books of Secrets in Medieval and Early Modern Culture* (Princeton, NJ: Princeton University Press, 1994).

Easton, S., *Roger Bacon and His Search for a Universal Science: A Reconsideration of the Life and Work of Roger Bacon in the Light of His Own Stated Purposes* (Oxford: Basil Blackwell, 1952).

Ebin, L., *Illuminator, Makar, Vates: Visions of Poetry in the Fifteenth Century* (Lincoln, NE: University of Nebraska Press, 1988).

Eco, U., *Art and Beauty in the Middle Ages*, trans. Hugh Bredin (New Haven & London: Yale University Press, 1986).

Elderedge, L. M., ed., *Benvenutus Grassus: The Wonderful Art of the Eye: A Critical Edition of De probatissima arte oculorum* (East Lansing, MI: Michigan State University Press, 1996).

Eleen, L., *The Illustration of the Pauline Epistles in French and English Bibles of the Twelfth and Thirteenth Centuries* (Oxford: Clarendon Press, 1982).

Emden, A. B., *A Biographical Register of the University of Oxford to A. D. 1500*, 3 vols (Oxford: Oxford University Press, 1957–59).

Evans, J., *Magical Jewels of the Middle Ages and the Renaissance* (Oxford: Clarendon Press, 1922; repr. New York: Dover, 1976).

Festal Menaion, trans. Mother Mary and K. Ware (South Canaan, PA: St Tikhon's Seminary Press, 1990).

Fletcher, A. J., *Preaching, Politics and Poetry in Late-Medieval England* (Dublin: Four Courts Press, 1998).

Flint, V. I. J., *The Rise of Magic in Early Medieval Europe* (Oxford: Clarendon Press, 1991).

Förster, M., 'Die altenglischen Bekenntnisformeln', *Englische Studien*, 75 (1942), 159–169.

Ford, J. A., *John Mirk's Festial: Orthodoxy, Lollardy, and the Common People in Fourteenth-Century England* (Cambridge: D. S. Brewer, 2006).

Fowler, R., ed., *Wulfstan's Canons of Edgar*, EETS, o.s. 266 (London, 1972).

Francis, W. N., ed., *The Book of Vices and Virtues*, EETS, o.s. 217 (London: Oxford University Press, 1942).

Frassetto, M., *Christian Attitudes towards the Jews in the Middle Ages* (New York–London: Routledge, 2007).

Fry, T., ed., *The Rule of St Benedict in Latin and English* (Minnesota: Liturgical Press, 1981).

Fulk, R. D., Robert E. Bjork, and John D. Niles, eds, *Klaeber's Beowulf and The Fight at Finnsburg* (Toronto: University of Toronto Press, 2008).

Furnivall, F. J., ed., *A Six-Text Print of ... Canterbury Tales*, part II, Chaucer Society, 1st series, XIV (London, 1870).

Furnivall, F. J., ed., *A Six-Text Print of Chaucer's Canterbury Tales*, part VIII, Chaucer Society, 1st series, XLIX (London, 1877).

Furnivall, F. J. and I. Gollancz, ed., *Hoccleve's Works: The Minor Poems including The Regement of Princes*, EETS, e.s. 72 (London: Keegan Paul, Trench & Trübner, 1897).

Gage, J., *Colour and Culture: Practice and Meaning from Antiquity to Abstraction* (London: Thames and Hudson, 1993).

Gerald of Wales, *The Jewel of the Church: A Translation of Gemma Ecclesiastica by Giraldus Cambrensis*, trans. J. Hagen (Leiden: Brill, 1979).

Geremek, B., *La stirpe di Caino: l'immagine dei vagabondi e dei poveri nelle letterature europee dal XV al XVII secolo*, ed. F. M. Cataluccio (Milano: Il saggiatore, 1988).
Getz, F. M., ed., *Healing and Society in Medieval England: A Middle English Translation of the Pharmaceutical Writings of Gilbertus Anglicus* (Madison, WI: University of Wisconsin Press, 1991).
Getz, F. M., 'The Faculty of Medicine before 1500', in *The History of the University of Oxford, Volume II: Late Medieval Oxford*, ed. Jeremy I. Catto and Ralph Evans (Oxford: Oxford University Press, 1992), pp. 373–413.
Gifford, D. J., 'Iconographical Notes towards a Definition of the Medieval Fool', *Journal of the Warburg and Courtauld Institutes*, 37 (1974), 336–43.
Gilson, S. A., *Medieval Optics and Theories of Light in the Works of Dante* (Lewiston, NY: Mellen, 2000).
Glass, F. D., '"Then the eyes of the blind shall be opened": A Lintel from San Cassiano a Settimo', in *Reading Medieval Images: The Art Historian and the Object*, ed. E. Sears and Th. K. Thomas (Ann Arbor, MI: University of Michigan Press, 2002), pp. 142–50.
Gneuss, H., ed., *Hymnar und Hymnen im englischen Mittelalter*, Buchreihe der Anglia, 12 (Tübingen, Max Niemeyer, 1968).
Goethe, Johann Wolfgang von, *Goethe's Werke: Vollständige Ausgabe letzter Hand*, 60 vols (Stuttgart & Tübingen: Cotta, 1828–1842).
Gray, D., ed., *The Oxford Book of Late Medieval Verse and Prose* (Oxford: Oxford University Press, 1985).
Greenway, D. and J. Sayers, eds, *Chronicles of the Abbey of Bury St Edmunds* (Oxford: Oxford University Press, 1989).
Grein, C. W. M., ed., *Sprachschatz der angelsächsischen Dichter*, 2 vols, Bibliothek der angelsächsischen Poesie, III, 1 and 2 (Cassel and Göttingen: Georg H. Wigand, 1861, 1864), rev. J. J. Köhler and F. Holthausen (Heidelberg: Carl Winter: 1912–1914).
Grimbert, J. T., 'Effects of *Clair-Obscur* in *Le Bel Inconnu*', in *Courtly Literature: Culture and Context*, ed. Keith Busby and Erik Kooper (Amsterdam: J. Benjamins 1990), pp. 249–60.
Grosseteste, Robert, *On Light (De luce)*, trans. Clare Riedl (Milwaukee, Wis.: Marquette University Press, 1942).
Grosseteste, Robert, *Hexaemeron*, trans. C. F. J. Martin (Oxford: Published for the British Academy by Oxford University Press, 1996).
Gurevič, A. J., *Le categorie della cultura medievale*, trans. Clara Castelli (Torino: G. Einaudi, 1983).
Hackett, J., ed., *Roger Bacon and the Sciences* (Leiden: E.J. Brill, 1997).

Hackett, J., 'Experience and Demonstration in Roger Bacon: A Critical Review of some Modern Interpretations', in *Erfahrung und Beweis: Die Wissenschaften von der Natur im 13. und 14. Jahrhundert / Experience and Demonstration: The Sciences of Nature in the 13th and 14th Centuries*, ed. Alexander Fidora and Matthias Lutz-Bachmann (Berlin: Akademie Verlag, 2006), pp. 41–58.

Hahn, C., '*Visio Dei*: Changes in Medieval Visuality', in *Visuality Before and Beyond the Renaissance*, ed. R. S. Nelson, pp. 169–196.

Hall, J., *Dictionary of Subjects and Symbols in Art* (London: John Murray, 1974).

Harvey, E. R., *The Inward Wits: Psychological Theory in the Middle Ages and Renaissance* (London: Warburg Institute, University of London, 1975).

Harris, A., 'A Romanesque Candlestick in London', *Journal of the British Archaeological Association*, 27 (1964), 32–62.

Haseloff, G., *Die Psalterillustration im 13. Jahrhundert: Studien zur Geschichte der Buchmalerei in England, Frankreich und den Niederlanden* (Kiel: Selbstverlag des Verfassers, 1938).

Hastings, A., A. Mason and H. Pyper, *The Oxford Companion to Christian Thought* (Oxford: Oxford University Press, 2000).

Hatt, C. A., ed., *The English Works of John Fisher* (Oxford: Oxford University Press, 2002).

Hausherr, I., 'La Méthode d'oraison hésychaste', *Orientalia Christiana Analecta* 9 (1927), 100–210.

Hawkins, J., 'Seeing the Light? Blindness and Sanctity in Later Medieval England', in *Saints and Sanctity*, ed. Peter Clarke and Tony Claydon, Studies in Church History, 47 (Woodbridge: The Ecclesiastical History Society by the Boydell Press, 2011), pp. 148–58.

Hawkins, J., 'The Blind in Later Medieval England: Medical, Social and Religious Responses' (unpublished PhD thesis, University of East Anglia, 2011).

Heck, E., *Roger Bacon: Ein Mittelalterlicher Versuch Einer Historischen und Systematischen Religionswissenschaft* (Bonn: H. Bouvier u. Co., 1957).

Henderson, J., ed. and trans., *Aristophanes*, III, Loeb Classical Library (Cambridge, Massachusetts: Harvard University Press, 2000).

Hildegard of Bingen, *On Natural Philosophy and Medicine: Selections from Cause et cure*, trans. Margaret Berger (Cambridge: Brewer, 1999).

Hinman, C., ed., *The Norton Facsimile: The First Folio of Shakespeare* (New York: W. W. Norton & Company, 1968).

Hodgson, P., and G. M. Liegey, eds., *The Orcherd of Syon*, EETS, o.s. 258 (London: Oxford University Press, 1966).

Horner, P. J., *A Macaronic Sermon Collection from Late Medieval England: Oxford, MS. Bodley 649* (Toronto: Pontifical Institute of Mediaeval Studies, 2006).
Hudson, A., ed., *English Wycliffite Sermons*, 5 vols (Oxford: Clarendon Press, 1983–1996).
Hunt, H., *Joy-Bearing Grief: Tears of Contrition in the Writings of the Early Syrian and Byzantine Fathers* (Leiden: Brill, 2004).
Hunt, T., *Popular Medicine in Thirteenth Century England: Introduction and Texts* (Cambridge: Brewer, 1990).
Irwin, E., *Color Terms in Greek Poetry* (Toronto: Hakkert, 1974).
Jauss, H. R., *Toward an Aesthetic of Reception*, trans. Timothy Bahti (Brighton: Harvester, 1982).
Jones, E., ed., *Medieval Heraldry: Some Fourteenth Century Heraldic Works* (Cardiff: William Lewis, 1943).
Jones, P. M., *Medieval Medical Miniatures* (London: British Library in association with the Wellcome Institute for the History of Medicine, 1984).
Jost, K., ed., *Die 'Institutes of Polity, Civil and Ecclesiastical'*, Swiss Studies in English, 47 (Berne: Francke, 1959).
Jusserand, J. J., *English Wayfaring Life in the Middle Ages* (New York: Ernest Benn Ltd., 1950).
Kazhdan, A. P., ed., *Oxford Dictionary of Byzantium*, 3 vols (Oxford: Oxford University Press, 1991).
Kemmler, F., *Exempla in Context: A Historical and Critical Study of Robert Mannying of Brunne's Handlyng Synne* (Tübingen: Narr, 1984).
Kempis, Thomas à, *The Imitation of Christ: The First English Translation of the 'Imitatio Christi'*, ed., B. J. H. Biggs, EETS, 309 (Oxford: Oxford University Press, 1997).
Kelly, J. N. D., *Early Christian Creeds*, 3rd edn (London: Longman, 1972).
Kendrick, T. D., T. J. Brown, R. L. S. Bruce-Mitford et al., (eds), *Evangeliorum Quattuor Codex Lindisfarnensis*, 2 vols (Olten and Lausanne: Urs Graf, 1956, 1960).
Klaeber, F., ed., *The Later Genesis*, Englische Textbibliothek, new edn (Heidelberg: Carl Winter, 1931).
Klaeber, F., ed., *Beowulf and The Fight at Finnsburg*, 3rd edn (Boston, Massachusetts, later issues Lexington, Massachusetts, 1950).
Kock, E. A., *Plain Points and Puzzles: 60 Notes on Old English Poetry*, Lunds Universitets Årsskrift, n.s., class 1, vol. 17, nr. 7 (Lund: C. W. K. Gleerup, 1922).
Koechlin, R., *Les ivoires gothiques français*, 3 vols (Paris: F. De Nobele, 1968).

Krapp, G. and E. Van Kirk Dobbie, eds, *The Anglo-Saxon Poetic Records*, 6 vols (New York: Columbia University Press, 1931–1953).

Krivocheine, B., *In the Light of Christ* (Crestwood, NY: St Vladimir's Seminary Press, 1986).

Lalleman, P. J., 'Healing by a Mere Touch as a Christian Concept', *Tyndale Bulletin*, 48.2 (1997), 355–361.

Landsberg, S., *Medieval Gardens* (London: British Museum Press, 2002).

Langland, W., *The Vision of Piers Plowman*, ed. A. V. C. Schmidt (London: Dent, 1995).

Lathan, R. E., *Revised Medieval Latin Word-List from British and Irish Sources with Supplement* (London: Published for the British Academy by Oxford University Press, 1965).

Lauritis, J. A., gen. ed., Ralph A. Klinefelter and Vernon F. Gallagher, eds., *A Critical Edition of John Lydgate's Life of Our Lady* (Pittsburg, PA: Duquesne University, 1961).

Lawton, D., 'Dullness and the Fifteenth Century', *ELH*, 54 (1987), 761–799.

Leader, D. R., *A History of the University of Cambridge, Volume 1: The University to 1546* (Cambridge: Cambridge University Press, 1988).

Le Goff, J., 'L'exemplum et la rhétorique de la prédication aux XIIIe et XIVe siècles', in *Retorica e poetica tra i secoli XII e XIV*, ed. Claudio Leonardi and Enrico Menestò (Firenze: La Nuova Italia, 1988).

LeJeune, A., *Euclide et Ptolémée: Deux stades de l'optique géométrique grecque* (Louvain: Bibliothèque de l'Université, Bureaux du "Recueil", 1948).

LeJeune A., ed., *L'optique de Claude Ptolémée dans la version latine d'après l'arabe de l'émir Eugène de Sicile*, new edn (Leiden: Brill, 1989).

Lentes, T., '"As far as I can see...": Rituals of Gazing in the Late Middle Ages', in *The Mind's Eye: Art and Theological Argument in the Middle Ages*, ed. J. Hamburger and A.-M. Bouché (Princeton: Princeton University Press, 2004), pp. 360–73.

Lewis, C. S., *The Allegory of Love: A Study in Medieval Tradition* (Oxford: Oxford University Press, 1936).

Liebermann, F., ed., *Gesetze der Angelsachsen*, 3 vols (Halle: Max Niemeyer, 1898–1916).

Lindberg, D. C., *Roger Bacon's Philosophy of Nature: A Critical Edition, with English Translation, Introduction, and Notes of De multiplicatione specierum and De speculis comburentibus* (Oxford: Clarendon Press, 1983).

Lindberg, D. C., *Studies in the History of Medieval Optics* (London: Variorum, 1983).

Lindberg, D. C., *Roger Bacon and the Origins of Perspectiva in the Middle Ages: A Critical Edition and English Translation of Bacon's 'Perspectiva' with Introduction and Notes* (Oxford: Clarendon Press, 1996).
Lindberg, D. C., 'The Medieval Church Encounters the Classical Tradition: Saint Augustine, Roger Bacon, and the Handmaiden Metaphor', in *When Science and Christianity Meet*, ed. David C. Lindberg and Ronald Numbers (Chicago: University of Chicago Press, 2003), pp. 7–32.
Lindberg, D. C., *Theories of Vision from Al-Kindi to Kepler* (Chicago: University of Chicago Press, 1976).
Lindberg, D. C., 'The Science of Optics', in *Science in the Middle Ages*, ed. David C. Lindberg (Chicago: University of Chicago Press, 1978), pp. 338–368.
Lindberg, D. C., *The Beginnings of Western Science: The European Scientific Tradition in Philosophical, Religious, and Institutional Context, 600 BC — AD 1450* (Chicago: University of Chicago Press, 1992).
Lorris, Guillaume de, and Jean de Meun, *Le Roman de la Rose*, ed. Armand Strubel (Paris: Librairie générale française, 1992); trans. Frances Horgan (Oxford: Oxford University Press, 1994).
Lowes, J. L., 'The Loveres Malady of Hereos', *Modern Philology*, 11 (1914), 419–546.
Lubac, H. de, *Medieval Exegesis*, I: *The Four Senses of Scripture*, trans. Mark Sebanc (Edinburgh: T&T Clark, 1998).
Lucretius, *De rerum natura* in Loeb Classical Library, ed. W. H. D. Rouse and rev. Martin Ferguson Smith, Loeb Classical Library (Cambridge, MA: Harvard University Press, 1975).
Pseudo-Macarius, *The Fifty Spiritual Homilies and the Great Letter*, trans. G. Maloney (New York: Paulist Press, 1992).
Magennis, H., 'Imagery of Light in Old English Poetry: Traditions and Appropriations', *Anglia*, 125 (2007), 181–204.
Manning, Lord (Bernard), *The People's Faith in the Time of Wyclif* (Cambridge: Cambridge University Press, 1919).
Mannyng, Robert, *Robert of Brunne's 'Handlyng Synne'*, ed. Frederick J. Furnivall, EETS, o.s. 119, 123 (London: Oxford University Press, 1901).
Manzalaoui, M. A., ed., *Secretum secretorum: Nine English Versions*, EETS, e.s. 276 (Oxford: Oxford University Press, 1977).
Marquardt, H., *Die altenglischen Kenningar*, Schriften der Königsberger Gelehrten Gesellschaft, 14th year, Geisteswissenschaftliche Klasse, 3 (Halle: Max Niemeyer, 1938).
Maximus Confessor, *Selected Writings*, trans. G. Berthold (New York: Paulist Press, 1985).

Maxwell-Stuart, P. G., *Studies in Greek Colour Terminology*, 2 vols (Amsterdam: E. J. Brill, 1981).
Mayor, J. E. B., ed., *The English Works of John Fisher*, EETS, e.s. 27 (London: N. Trübner & Co, 1876).
McEvoy, J., *Robert Grosseteste* (Oxford: Oxford University Press, 2000).
McGuckin, J. A., 'Symeon the New Theologian's Hymns of Divine Eros: A Neglected Masterpiece of the Christian Mystical Tradition', *Spiritus: A Journal of Christian Spirituality*, 5.2 (2005), 182–202.
McVaugh, M., 'The *humidum radicale* in Thirteenth-century Medicine', *Traditio*, 30 (1974), 259–283.
McVaugh, M., 'An Early Discussion of Medical Degrees at Montpellier by Henry of Winchester', *Bulletin of the History of Medicine*, 49 (1975), 51–71.
Mellone, A., 'luce', in *Enciclopedia dantesca*, gen. ed. Umberto Bosco, 6 vols (Rome: Istituto delle Enciclopedia italiana, fondata da Giovanni Treccani, 1970–1976; 2nd edn 1984), III, p. 708.
Mellows, W. T., P. I. King and C. N. L. Brooke, eds, *The Book of William Morton, Almoner of Peterborough Monastery*, Northamptonshire Record Society, 16 (Oxford: printed for the Northamptonshire Record Society by Oxford University Press, 1954).
Ménard, P., 'Les illustrations marginales du "Roman d'Alexandre" (Oxford, Bodleian Library, Bodley 264)', in *Risus Mediaevalis: Laughter in Medieval Literature and Art* (Leuven: Leuven University Press, 2003), pp. 75–118.
Merback, M. B., *The Thief, the Cross and the Wheel: Pain and the Spectacle of Punishment in Medieval and Renaissance Europe* (London: Reaktion Books, 1999).
Merwe Scholtz, H. van der, *The Kenning in Anglo-Saxon and Old Norse Poetry* (Utrecht doctoral dissertation, 1927).
Meyer, R. M., *Die altgermanische Poesie nach ihren formelhaften Elementen beschrieben* (Berlin: Wilhelm Hertz, 1889).
Miles, M. R., 'Vision: The Eye of the Body and the Eye of the Mind in Saint Augustine's *De trinitate* and *Confessions*', *The Journal of Religion*, 63 (1983), 125–142.
Milfull, I. B., ed., *The Hymns of the Anglo-Saxon Church* (Cambridge: University Press, 1996).
Millet, B., ed., *Ancrene Wisse: A Corrected Edition of the Text in Cambridge, Corpus Christi College, MS 402*, EETS, o.s. 325, 2 vols (Oxford: Oxford University Press, 2005).
Mirecki, P. and J. BeDuhn, eds, *The Light and the Darkness: Studies in Manichaeism and its World* (Leiden: Brill, 2001).

Mirk, John, *Mirk's Festial: A Collection of Homilies*, ed. Theodor Erbe, EETS, o.s. 96 (London, Keegan Paul, Trench & Trübner, 1936).
Mitchell, J. A., 'Queen Katherine and the Secret of Lydgate's Temple of Glass', *Medium Ævum*, 77 (2008), 53–76.
Mollat, M., *Les pauvres au Moyen Âge: Étude sociale* (Paris: Hachette, 1978).
Morray-Jones, C. R. A., 'Paradise Revisited (2 Cor 12: 1–12): The Jewish Mystical Background of Paul's Apostolate. Part 2: Paul's Heavenly Ascent and Its Significance', *Harvard Theological Review*, 86 (1993), 265–92.
Morrill, G. L., ed., *Speculum Guy de Warewyke*, EETS, e.s. 75 (London: Keegan Paul, Trench & Trübner, 1898).
Mortimer, N., *John Lydgate's Fall of Princes: Narrative Tragedy in its Literary and Political Contexts* (Oxford: Clarendon Press, 2005).
Myrc, John, *Instructions for Parish Priests*, ed. E. Peacock, EETS, o.s. 31 (London: Keegan Paul, Trench & Trübner, 1902).
Neale, J. M. and R. F. Littledale, *A Commentary on the Psalms from Primitive and Mediæval Writers*, 4th edn, 4 vols (London: Joseph Masters & Co., 1884).
Nelson R. S., ed., *Visuality Before and Beyond the Renaissance: Seeing as Others Saw* (Cambridge: Cambridge University Press, 2000).
Nevalinna, S., ed., *The Northern Homily Cycle: The Expanded Version in MSS Harley 4196 and Cotton Tiberius E vii III*, 3 vols (Helsinki: Mémoires de la Société Néophilologique, 1972–1984).
Newman, J. H. and M. Pattison, trans., *Catena Aurea ... collected by S. Thomas Aquinas*, 4 vols in 8 parts (Oxford: John Henry Parker, 1841–1845).
Nutton, V., 'Medicine in the Greek World, 800–50 BC', in *The Western Medical Tradition: 800 BC to AD 1800*, ed. Lawrence I. Conrad et al. (Cambridge: Cambridge University Press, 1995).
Ogden, M. S., ed., *Cyrurgie of Guy de Chauliac*, EETS, o.s. 265 (Oxford: Oxford University Press, 1971).
Oliver, J. H., *Gothic Manuscript Illumination in the Diocese of Liège (c.1250–c.1330)* (Leuven: Uitgeverij Peeters, 1988).
Oman, C., 'The Gloucester Candlestick', in *Victoria and Albert Museum Monographs*, 11 (1958), 1–14.
Palmer, R., 'In Bad Odour: Smell and its Significance in Medicine from Antiquity to the Seventeenth Century, in *Medicine and the Five Senses*, ed. W. F. Bynum and Roy Porter (Cambridge: Cambridge University Press, 1993), pp. 61–68.
Pastoureau, M., 'L'Eglise et la couleur, des origines à la Réforme', *Bibliothèque de l'École des chartes*, 147 (1989), 203–230.

Patterson, L., 'Perpetual Motion: Alchemy and the Technology of the Self', in id., *Temporal Circumstances: Form and History in the Canterbury Tales* (New York: Palgrave, 2006), pp. 159–176.
Pearsall, D., *John Lydgate* (London: Routledge and Kegan Paul, 1970).
Pearsall, D., *John Lydgate (1371–1449): A Bio-bibliography* (Victoria, BC: University of Victoria, 1997).
Petrina, A., *Cultural Politics in Fifteenth-Century England: The Case of Humphrey, Duke of Gloucester* (Leiden: Brill, 2004).
Petrus Hyspanus (Pope John XXI), *The Treasury of Health* (London: Wyllyam Coplande, [1550?]).
Pinborg, Jan, 'Roger Bacon on Signs: A Newly Recovered Part of the Opus maius', in Jan P. Beckmann et al., eds, *Sprache und Erkenntnis im Mittelalter: Akten des VI. Internationalen Kongress für Mittelalterliche Philosophie der la Société internationale pour l'étude de la philosophie médiévale, 29 August-3 Septembre 1977* (Berlin: Walter de Gruyter, 1981).
Plato, *Timaeus*, trans. Donald J. Zeyl (Cambridge/Indianapolis: Hackett, 2000).
Pleij, H., *Colors Demonic and Divine: Shades of Meaning in the Middle Ages and After*, trans. Diane Webb (New York: Columbia University Press, 2002, 2004).
Pliny, *Natural History*, trans. H. Rackham, 10 vols (Cambridge, MA: Harvard University Press, 1952; repr. 1961), vol. IX.
Pokorny, J., ed., *Indogermanisches etymologisches Wörterbuch*, 2 vols (Berne: Francke, 1959–1969).
Postles, D., 'Lights, Lamps and Layfolk: "Popular" Devotion Before the Black Death', *Journal of Medieval History*, 25 (1999), 97–114.
Pouchelle, M.-C., *The Body and Surgery in the Middle Ages*, trans. Rosemary Morris (New Brunswick, NJ: Rutgers University Press, 1990).
Powell, M. G., *Introduction to Medieval Christian Liturgy* <http://www.yale.edu/adhoc/research_resources/liturgy/hours.html>.
Powell, S., ed., *John Mirk's 'Festial' edited from British Library MS Cotton Claudius A. II*, vol. I, EETS, o.s. 334 (2009).
Power, A., 'A Mirror for Every Age: The Reputation of Roger Bacon', *English Historical Review* 121 (2006), 657–92.
Power, A., *Roger Bacon and the Defence of Christendom* (Cambridge: Cambridge University Press, 2012).
Powicke, F. M. and C. R. Cheney, eds, *Councils and Synods with Other Documents Relating to the English Church*, 2 vols (Oxford: Clarendon Press, 1964).

Raine, J., and J. W. Clay, eds, *Testamenta Eboracensia: A Selection of Wills from the Registry at York (1300–1551)*, vol. II, Surtees Society, 45 (Durham: published for the Society by George Andrews & Co, 1864).

Randall, L. M. C., *Images in the Margins of Gothic Manuscripts* (Berkeley–Los Angeles: University of California Press, 1966).

Randall, R. H., 'Frog in the Middle', *The Metropolitan Museum of Art Bulletin*, 16 (1958), 269–275.

Rawcliffe, C., *Leprosy in Medieval England* (Woodbridge: Boydell, 2006).

Rawcliffe, C., '"Delectable Sightes and Fragrant Smells"': Gardens and Health in Late Medieval and Early Modern England', *Garden History*, 36 (2008), 3–21.

Raymond, D. R., ed., *Dispatches from the Front: The* Prefaces *to the Oxford English Dictionary* (Waterloo, Ontario: Centre for the New Oxford English Dictionary, University of Waterloo, 1987).

Renoir, A., *The Poetry of John Lydgate* (London: Routledge and Kegan Paul, 1967).

Rhodes, M., 'A Pair of Fifteenth-Century Spectacle Frames from the City of London', *Antiquaries*, 62 (1982), 57–73.

Ribémont, B., 'On the Definition of an Encyclopædic Genre in the Middle Ages', in *Pre-Modern Encyclopædic Texts: Proceedings of the Second COMERS Congress, Groningen, 1–4 July 1996*, ed. Peter Brinkley (Leiden: Brill, 1997), pp. 46–61.

Riet, S. van, ed., *Avicenna Latinus: Liber de anima seu sextus de naturalibus*, I–II–III (Louvain and Leiden: E.J. Brill and Peeters, 1972).

Ritson, J., *Bibliographia Poetica: A catalogue of English poets of the twelfth, thirteenth, fourteenth, fifteenth, and sixteenth centurys, with a short account of their works* (London: Roworth, 1802).

Roberts, P. B., 'Preaching in / and the Medieval City', in *Medieval Sermons and Society: Cloister, City, University*, ed. Jacqueline Hamesse et al. (Louvain-la-Neuve: Fédération internationale des instituts d'études médiévales, 1998).

Robinson, P., intro., *Manuscript Tanner 346: A Facsimile*, The Facsimile Series of the Works of Geoffrey Chaucer, vol. I (Norman, OK: Pilgrim, 1980).

Ross, A. S. C., 'OE. "leoht" 'world"', *Notes and Queries*, 220 (1975), 196.

Rossiter, W. T., *Chaucer and Petrarch* (Woodbridge and Rochester, NY: Boydell and Brewer, 2010).

Rossiter, W. T., '"Disgraces the name and patronage of his master Chaucer": Echoes and Reflections in Lydgate's Courtly Poetry', in *Standing in the Shadow of the Master: Chaucerian Influences and Interpretations*, ed. Kathleen A. Bishop (Newcastle: Cambridge Scholars Press, 2010), pp. 2–27.

Rossiter, W. T., 'The Marginalization of John Lydgate', *Marginalia*, 1 (2005) <http://marginalia.co.uk/journal/05margins/rossiter.php>.

Rundle, D., ed., *Humanism in Fifteenth-Century Europe* (Oxford: Society for the Study of Medieval Languages and Literature, 2012).

Samson, G. R., 'Historical Trends in the Deployment of Tinctures', *The Coat of Arms* n.s., 13 (2000), 271–7.

Sand, A., 'Vision, Devotion, and Difficulty in the Psalter Hours "Of Yolande of Soissons"', *The Art Bulletin*, 87.1 (2005), 6–23.

Sandler, L. F., *Gothic Manuscripts 1285–1385: A Survey of Manuscripts Illuminated in the British Isles* (London – Oxford: Harvey Miller Publishers – Oxford University Press, 1986).

Sandler, L. F., *Omne bonum: A Fourteenth-Century Encyclopedia of Universal Knowledge. British Library MSS Royal 6 E VI – 6 E VII* (London: Harvey Miller Publishers, 1996).

Sandler, L. F., 'Pictorial and Verbal Play in the Margins: The Case of British Library, Stowe MS 49', in *Illuminating The Book: Makers and Interpreters. Essays in Honour of Janet Backhouse*, ed. Michelle P. Brown and Scot McKendrick (London: The British Library, 1998), pp. 52–68.

Sandler, L. F., 'The Images of Words in English Gothic Psalters', in *Studies of the Illustration of the Psalter*, ed. B. Cassidy and R. Muir Wright (Stamford: Shaun Tyas, 2000), pp. 52–68.

Sauer, H., ed., *Theodulfi Capitula in England*, Münchener Universitäts-Schriften, Texte und Untersuchungen zur Englischen Philologie, 8 (Munich: Wilhelm Fink, 1978).

Scanlon, L., *Narrative, Authority and Power: The Medieval Exemplum and the Chaucerian Tradition* (Cambridge: Cambridge University Press, 1994).

Scanlon, L., 'Lydgate's Poetics: Laureation and Domesticity in the *Temple of Glass*', in *John Lydgate: Poetry, Culture, and Lancastrian England*, ed. Larry Scanlon and James Simpson (Notre Dame, IN: University of Notre Dame Press, 2006), pp. 61–97.

Schick, J., ed., *Lydgate's Temple of Glas*, EETS, e.s. 60 (London: Kegan Paul, Trench, Trübner, 1891).

Schless, H. H., *Chaucer and Dante: A Revaluation* (Norman, OK: Pilgrim, 1984).

Sehrt, E. H., ed., *Vollständiges Wörterbuch zum Heliand und zur altsächsischen Genesis*, Hesperia, XIV (1925, repr. Göttingen: Vandenhoeck & Ruprecht, 1966).

Seneca [the Younger], *Naturales Quaestiones I*, trans. T. H. Corcoran, Seneca in Ten Volumes (Cambridge, MA: Harvard University Press, 1971), vol. VII.

Seymour, M. C., ed., *On the Properties of Things: John Trevisa's Translation of Bartholomæus Anglicus De Proprietatibus Rerum. A Critical Text*, 3 vols (Oxford: Clarendon Press, 1975–1988).

Sharpe, R. R., ed., *Calendar of Wills Proved and Enrolled in the Court of Husting, London, 1258–1688*, 2 vols (London: Printed by John C. Francis, Took's Court, 1889–1890).

Siegel, R. E., ed., *Galen on Sense Perception* (Basel: Karger, 1970).

Sieper, E., ed., *Lydgate's Reson and Sensuallyte*, 2 vols, EETS, e.s. 84, 89 (London: Kegan Paul, Trench, Trübner, 1901–1903).

Sievers, E., ed., *Der Heliand und die angelsächsische Genesis* (Halle: Lippert'sche Buchhandlung, Max Niemeyer, 1875).

Sievers, E., ed., *Heliand* (Halle: Buchhandlung des Waisenhauses, 1878; reissued with appendix by Edward Schröder, 1935).

Simpson, J., *Reform and Cultural Revolution*, The Oxford English Literary History, volume 2: 1350–1547 (Oxford: Oxford University Press, 2002).

Simpson, J., *Piers Plowman: An Introduction to the B-Text* (London: Longman, 1990; rev. ed. Exeter: University of Exeter Press, 2007).

Singelenberg, P., 'The Iconography of the Etschmiadzin Diptych and the Healing of the Blind Man at Siloe', *The Art Bulletin*, 40.2 (1958), 105–112.

Siraisi, N., *Medieval and Early Renaissance Medicine: An Introduction to Knowledge and Practice* (Chicago: University of Chicago Press, 1990).

Smalley, B., *The Study of the Bible in the Middle Ages* (Notre Dame, IN: Notre Dame University Press, 1978).

Smith, A. M., *Ptolemy's Theory of Visual Perception: An English Translation of the Optics with Introduction and Commentary*, (Philadelphia: The American Philosophical Society, 1996 [=*Transactions of the American Philosophical Society* 86, part 2]).

Sommer, H., ed., *The Recuyell of the Historyes of Troye*, trans. William Caxton (London: D. Nutt, 1892).

Sotres, P. G., 'The Regimens of Health', in *Western Medical Thought from Antiquity to the Middle Ages*, ed. M. D. Grmek, trans. Anthony Shugaar (Cambridge, MA: Harvard University Press, 1998).

Spearing, A. C., *Medieval Dream-Poetry* (Cambridge: Cambridge University Press, 1976).

Spearing, A. C., *The Medieval Poet as Voyeur* (Cambridge: Cambridge University Press, 1993).

Spencer, H. L., *English Preaching in the Late Middle Ages* (Oxford: Clarendon Press, 1993).

Stanley, E. G., 'Polysemy and Synonymy and How these Concepts were Understood from the Eighteenth Century Onwards in Treatises, and

Applied in Dictionaries of English', in Julie Coleman and Anne McDermott, eds, *Historical Dictionaries and Historical Dictionary Research*, Lexicographica Series Maior, 123 (Tübingen: Max Niemeyer, 2004), pp. 157–83.

Steele, R., *Medieval Lore from Bartholomew Anglicus* (London: Alexander Moring, The De La More Press, 1905).

Steele, R., ed., *Lydgate and Burgh's Secrees of Old Philisoffres*, EETS, e.s. 66 (London: Keegan Paul, Trench & Trübner, 1894).

Steele, R., ed., *Three Prose Versions of the Secreta Secretorum*, EETS, e.s. 74 (London: Keegan Paul, Trench & Trübner, 1898).

Stevenson, J., 'A New Type of Late Medieval Spectacle Frame from the City of London', *London Archaeologist*, 7 (1995), 321–27.

Stones, A., 'The Full-Page Miniatures of the Psalter-Hours New York, PML, MS M.729. Programme and Patron', in *The Illuminated Psalter: Studies in the Content, Purpose and Placement of its Images*, ed. F. O. Büttner (Turnhout: Brepols, 2004), pp. 281–307.

Strohm, P., 'Chaucer's Fifteenth Century Audience and the Narrowing of the "Chaucer Tradition"', *Studies in the Age of Chaucer*, 4 (1982), 3–32.

Southern, R. W., *Robert Grosseteste: The Growth of an English Mind in Medieval Europe* (Oxford: Clarendon Press, 1986).

Swanson, R. N., *Religion and Devotion in Europe, c.1215–c.1515* (Cambridge: Cambridge University Press, 1995).

Sydenham, C., 'Translating the Gloucester Candlestick', in *The Burlington Magazine*, 126.977 (1984), 504.

Symeon the New Theologian, *The Discourses*, trans. C. J. de Catanzaro (New York: Paulist Press, 1980).

Symeon the New Theologian, *The Practical and Theological Chapters and the Three Theological Discourses*, trans. P. (= J. A.) McGuckin (Kalamazoo, MI: Cistercian Publications, 1982).

Symeon the New Theologian, *Hymns of Divine Love*, intro. and trans. G. A. Maloney (Denville, NJ: Dimension Books, 1975).

Symeon the New Theologian, *On The Mystical Life*, vol. 2, *On Virtue and Christian Life*, trans. A. Golitzin (Crestwood, NY: St Vladimir's Seminary Press, 1996).

Tachau, K., *Vision and Certitude in the Age of Ockham: Optics, Epistemology and the Foundations of Semantics 1250–1345* (Leiden: E.J. Brill, 1988).

Tachau, K., 'Seeing as Action and Passion in the Thirteenth and Fourteenth Centuries', in *The Mind's Eye: Art and Theological Argument in the Middle Ages*, ed. J. Hamburger and A.-M. Bouché (Princeton: Princeton University Press, 2004), pp. 336–359.

Thompson, A. B., *The Northern Homily Cycle* (Kalamazoo, MI: Medieval Institute Publications, 2008).
Thurlby, M., *The Herefordshire School of Romanesque Sculpture* (Hereford: Logaston Press, 1999).
Timmer, B. J., ed., *The Later Genesis* (Oxford: Scrivener Press, 1948).
Torti, A., *The Glass of Form: Mirroring Structures from Chaucer to Skelton* (Cambridge: Brewer, 1991).
Treadgold, W., *A History of the Byzantine State and Society* (Stanford: Stanford University Press, 1997).
The Holie Bible Faithfully Translated into English ... By the English College of Doway, 2 vols (Doway: by Laurence Kellam, 1609, 1610).
The Holy Bible. An Exact Reprint in Roman Type ... of the Authorized Version Published in the Year 1611, intro. Alfred W. Pollard (London: Henry Frowde, Oxford University Press, 1911).
The New Testament of Iesus Christ, Translated Faithfully into English out of the authentical Latin ... in the English College of Rhemes (Rhemes: by Iohn Fogny, 1582)
The Text of the New Testament of Iesus Christ, Translated Out of the vulgar Latine by the Papists of the traiterous Seminarie at RHEMES ... with A Confutation of all such Arguments, Glosses, and Annotations, As Conteine Manifest impietie, of heresie and slander, against the Catholike Church of GOD... by William Fulke (London: by Christopher Barker, 1589)
Thomas Aquinas, *Sentencia libri De anima*, cura et studio Fratrum Praedicatorum, Opera Omnia, iussu Leonis XIII P.M. edita, Tomus XLV (Rome & Paris: Commissio Leonina & J. Vrin, 1984).
Thomas Aquinas, *Commentary on Aristotle's De anima*, trans. Kenelm Foster and Silvester Humphries, intro. Ralph McInerny, (Notre Dame, IN: Dumb Ox Books, 1994).
Thompson, C., *Alchemy and Alchemists* (New York: Dover, 2002).
Vogl, S., 'Roger Bacons Lehre von der sinnlichen Spezies und vom Sehvorgange', in *Roger Bacon*, ed. A. G. Little (Oxford: Clarendon Press, 1914).
Wakelin, D., *Humanism, Reading, and English Literature 1430–1530* (Oxford: Oxford University Press, 2007).
Walde, A., ed., *Lateinisches etymologisches Wörterbuch*, 3rd edn, rev. J. B. Hofmann, 3 vols (Heidelberg: Carl Winter, 1938–1956)
Warren, F. E., trans., *The Sarum Missal in English*, 2 vols, The Library of Liturgiology & Ecclesiology for English Readers, ed. Vernon Staley, VIII, IX (London: The De La More Press, 1911).

Weatherly, E. H., ed., *Speculum sacerdotale*, EETS, o.s. 200 (London: Oxford University Press, 1936).

Weiss, R., *Humanism in England during the Fifteenth Century*, 4th edn, ed. D. Rundle and A. J. Lappin (Oxford: Society for the Study of Medieval Languages and Literature, 2010).

Wenzel, S., *The Sin of Sloth: Acedia in Medieval Thought and Literature* (Chapel Hill: University of North Carolina Press, 1960).

Wenzel, S., *Macaronic Sermons: Bilingualism and Preaching in Late-Medieval England* (Ann Arbor, MI: University of Michigan Press, 1994).

Wheatley, E., 'The Blind Beating the Blind: An Unidentified "Game" in a Marginal Illustration of the Romance of Alexander, MS Bodley 264', *Journal of the Warburg and Courtauld Institutes*, 68 (2005), 213–217.

Wheatley, E., *Stumbling Blocks Before the Blind: Medieval Constructions of a Disability* (Ann Arbo, MI: University of Michigan Press, 2010).

White, H., trans., *Ancrene Wisse: Guide for Anchoresses* (Harmondsworth: Penguin, 1993).

Whitelock, D., ed., *Sweet's Anglo-Saxon Reader*, rev. edn (Oxford: Clarendon Press, 1967).

Wickham Legg, J., ed., *The Sarum Missal* (Oxford: Clarendon Press, 1916, repr. 1969).

Wordsworth, John, Henry Julian White, et al., eds, *Nouum Testamentum Domini Nostri Iesu Christi Latine*, 3 vols (Oxford: Clarendon Press, 1889–1954).

William of Conches, *Glosae super Platonem : texte critique avec notes et tables*, ed. Edouard Jeauneau, Textes Philosophiques du Moyen Âge XIII (Paris: J. Vrin, 1965).

Wilson, J., 'Poet and Patron in Early Fifteenth-century England: John Lydgate's Temple of Glas', *Parergon* 11 (1975), 25–32.

Wimsatt, J. I., *Chaucer and His Contemporaries: Natural Music in the Fourteenth Century* (Toronto: University of Toronto Press, 1991).

Youngs, D., and S. Harris, 'Demonizing the Night in Medieval Europe: A Temporal Monstrosity?', in *The Monstrous Middle Ages*, ed. Bettina Bildhauer and Robert Mills (Cardiff: University of Wales Press, 2003), pp. 134–154.

Zangemeister, K. and W. Braune, eds, 'Bruchstücke der altsächsischen Bibeldichtung aus der Bibliotheca Palatina', *Neue Heidelberger Jahrbücher*, 4 (1894), 205–94.

Zarnecki, G., J. Holt and T. Holland, eds, *English Romanesque Art 1066–1200* (London: Arts Council of Great Britain in association with Weidenfeld and Nicholson, 1984).

Zonno, S., 'Il mirabile castello: Davide e la sua corte nel Salterio inglese della Biblioteca Queriniana di Brescia', *Annali Queriniani*, 6 (2005), 71–100.

Zonno, S., 'Illumination Translates: The Image of the Castle in Some Fourteenth-Century English Manuscripts, in *The Medieval Translator-Traduire au Moyen Âge: Vol. 12 Lost in Translation*, ed. D. Renevey and C. Whitehead (Turnhout: Brepols, 2009), pp. 297–313.

www.ingramcontent.com/pod-product-compliance
Lightning Source LLC
Chambersburg PA
CBHW022016220426
43663CB00007B/1100